The Foundations of Statistics:
A Simulation-based Approach

Shravan Vasishth · Michael Broe

The Foundations of Statistics:
A Simulation-based Approach

 Springer

Shravan Vasishth
Department of Linguistics
University of Potsdam
Karl-Liebknecht-Str. 24-25
14476 Potsdam
Germany
vasishth@uni-potsdam.de

Michael Broe
Department of Evolution,
Ecology & Organismal Biology
Ohio State University
1304 Museum of Biological Diversity
Kinnear Road 1315
OH 43212 Columbus
USA
broe.1@osu.edu

ISBN 978-3-642-42344-4   ISBN 978-3-642-16313-5 (eBook)
DOI 10.1007/978-3-642-16313-5
Springer Heidelberg Dordrecht London New York

Cover design: WMXDesign GmbH

Cover image: Daniel A. Becker

Printed on acid-free paper

Springer is part of Springer Science+Business Media (www.springer.com)

SV dedicates this book to his son, Atri;
MB dedicates this book to his parents.

Foreword

Teaching the fundamental concepts and ideas of statistics is a relatively easy task when students have already completed courses in probability theory and calculus. Such students not only come well-prepared to an introductory statistics course, they are likely to follow additional, more advanced courses in mathematics and related areas. If so, many of the statistical techniques acquired remain relevant and are developed by further training.

For students without much of a background in probability or calculus, the situation is quite different. For many of us, just about everything is new, from integration to probability density functions, and from slopes of regression lines to random variables. Traditionally, introductory statistics courses targeting students from the arts and social sciences seek to explain the basic concepts and results of mathematical statistics, while omitting proofs. To survive such a course, one typically dutifully memorizes the equation for the density of the normal curve, the definition of the central limit theorem, and the sums of squares of a one-way analysis of variance. Once through this bottleneck, there is the safe haven of menu-driven software packages that will calculate everything one needs (and more). Sadly enough, many students will not come to appreciate the beauty of the fundamentals of statistics, and they will also remain somewhat insecure about what the basic concepts actually mean.

The approach taken by Shravan Vasishth and Michael Broe provides a much more interesting, exciting, and I believe lasting learning experience for students in the social sciences and the arts. (In fact, I believe that students in the sciences will also enjoy lots of the R code in this book). Simulation is a wonderful way of demystifying concepts and techniques that would otherwise remain abstract and distant. By playing around with simulated data, the reader can quickly get an intuitive feel for basic concepts such as the sampling distribution of the sample mean. With the tools provided in this book, the reader can indeed begin to explore the foundations of statistics, and will discover that statistics is actually fun and rewarding. Along the way, the reader will also acquire some basic programming skills, and will be well prepared to use the R software environment for statistical computing and graphics. R is attractive not only because it is open source, and comes with thousands of add-on packages for specialized data analysis. R is also a great choice because of the many excellent books introducing statistical techniques in a reader-friendly and accessible way, with lots of example data sets. With this introduction, Shravan Vasishth and Michael Broe complement the existing R literature with a very useful and enjoyable hands-on introduction to statistics, doing analysis by synthesis.

Edmonton, October 2010   *R. Harald Baayen*

Preface

Statistics and hypothesis testing are routinely used in areas that are traditionally not mathematically demanding (an example is psycholinguistics). In such fields, when faced with experimental data in any form, many students and researchers tend to rely on commercial packages to carry out statistical data analysis, often without acquiring much understanding of the logic of statistics they rely on. There are two major problems with this approach. First, the results are often misinterpreted. Second, users are rarely able to flexibly apply techniques relevant to their own research – they use whatever they happened to have learnt from their advisors, and if a slightly new data analysis situation arises, they are unable to use a different method.

A simple solution to the first problem is to teach the foundational ideas of statistical hypothesis testing without using too much mathematics. In order to achieve this, statistics instructors routinely present simulations to students in order to help them intuitively understand things like the Central Limit Theorem. This approach appears to facilitate understanding, but this understanding is fleeting. A deeper and more permanent appreciation of the foundational ideas can be achieved if students re-run and modify the simulations themselves outside the class.

This book is an attempt to address the problem of superficial understanding. It provides a largely non-mathematical, simulation-based introduction to basic statistical concepts, and encourages the reader to try out the simulations themselves using the code provided on the course homepage http://www.purl.oclc.org/NET/vasishth/VB/. Since the exercises provided in the text almost always require the use of programming constructs previously introduced, the diligent student acquires basic programming ability as a side effect. This helps to build up the confidence necessary for carrying out more sophisticated analyses. The present book can be considered as the background material necessary for more advanced courses in statistics.

The vehicle for simulation is a freely available software package, R (see the CRAN website for further details). This book is written using Sweave (pronounced S-weave), which was developed by Leisch, 2002. This means that LaTeX and R code are interwoven together.

The style of presentation used in this book is based on a course developed by Michael Broe in the Linguistics department of The Ohio State University. The first author (SV) was a student at the time and attended Michael's course in 2000; later, SV extended the book in the spirit of the original course (which was prepared using commercially available software). Both authors collaborated on the final text.

SV has used this book to teach linguistics undergraduate and graduate students at the University of Saarland, the University of Potsdam, and at the European Summer Schools for Language, Logic and Information held in Edinburgh (2005) and Bordeaux (2009). These courses have shown that the highly motivated student with little to no programming ability and/or mathematical/statistical training can understand everything presented here, and can move on to using R and statistics productively and sensibly.

The book is designed for self-instruction or to accompany a statistics course that involves the use of computers. Some of the examples are from linguistics, but this does not affect the content, which is of general relevance to any scientific discipline. The reader will benefit, as we did, by working through the present book while also consulting some of the books we relied on, in particular Rietveld & van Hout, 2005; Maxwell & Delaney, 2000; Baayen, 2008; Gelman & Hill, 2007.

We do not aspire to teach R per se in this book; if this book is used for self-instruction, the reader is expected to either take the initiative themselves to acquire a basic understanding of R, and if this book is used in a taught course, the first few lectures should be devoted to a simple introduction to R.

After completing this book, the reader will be ready to use more advanced books like Gelman and Hill's *Data analysis using regression and multilevel/hierarchical models*), Baayen's *Analyzing Linguistic Data*, and the online lecture notes by Roger Levy.

A lot of people were directly or indirectly involved in the creation of this book. Thanks go to Olga Chiarcos and Federica Corradi dell Acqua at Springer for patiently putting up with delays in the preparation of this book. In particular, without Olga's initiative and efforts, this book would not have appeared in printed form. SV thanks Reinhold Kliegl, Professor of Psychology at the University of Potsdam, for generously sharing his insights into statistical theory and linear mixed models in particular, and for the opportunity to co-teach courses on statistical data analysis with him; his comments also significantly improved chapter 7. Harald Baayen carefully read the entire book and made important suggestions for improvement; our thanks to him for taking the time. Thanks also to the many students (among them: Pavel Logačev, Rukshin Shaher, and Titus von der Malsburg) who commented on earlier versions of this book. SV is grateful to Andrea Vasishth for support in every aspect of life.

MB thanks the students at The Ohio State University who participated in the development of the original course, and Mary Tait for continued faith and support.

We end with a note on the book cover; it shows a visualization of Galton's box (design by Daniel A. Becker). It is also known as the bean machine or quincunx. This was originally a mechanical device invented by Sir Francis Galton to demonstrate the normal distribution and the central limit theorem. The reader can play with the quincunx using the R version (written by Andrej Blejec) of the simulation, shown below (the code is available from Dr. Blejec's homepage and from the source code accompanying this book). The result of Dr. Blejec's code is shown below.

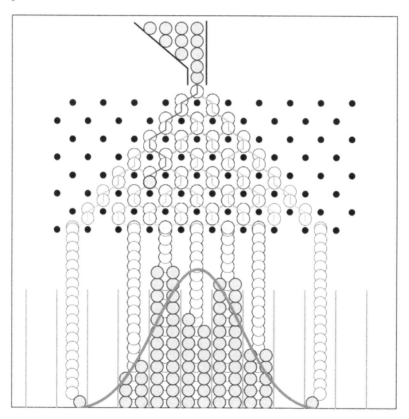

Berlin, Germany; and Columbus, OH   *Shravan Vasishth*
October 2010   *Michael Broe*

Contents

Chapter 1
Getting Started

The main goal of this book is to help you understand the principles behind inferential statistics, and to use and customize statistical tests to your needs. The vehicle for this will be a programming language called R (google for 'CRAN,' which stands for the Comprehensive R Archive Network). Let's start with installing R and related software.

1.1 Installation: R, LaTeX, and Emacs

In order to use the code that comes with this book, you only need to install R. The latest version can be downloaded from the CRAN website. However, other freely available software provides a set of tools that work together with R to give a very pleasant computing environment. The least that you need to know about is LaTeX, Emacs, and Emacs Speaks Statistics. Other tools that will further enhance your working experience with LaTeX are AucTeX, RefTeX, preview-latex, and Python. None of these are required but are highly recommended for typesetting and other sub-tasks necessary for data analysis.

There are many advantages to using R with these tools. For example, R and LaTeX code can be intermixed in Emacs using noweb mode. R can output data tables etc. in LaTeX format, allowing you to efficiently integrate your scientific writing with the data analysis. This book was typeset using many of the above tools. For more on LaTeX, see the http://www.tug.org, the TeX Users Group website.

The installation of this working environment differs from one operating system to another. In Linux-like environments, most of these tools are already pre-installed. For Windows and Macintosh you will need to read the manual pages on the CRAN website.

After you have installed R on your machine, the second thing you need to do before proceeding any further with this book is to learn a little bit about R. The present book is not intended to be an introduction to R. For short,

S. Vasishth, M. Broe, *The Foundations of Statistics: A Simulation-based Approach*, DOI 10.1007/978-3-642-16313-5_1,
© Springer-Verlag Berlin Heidelberg 2011

comprehensive and freely available introductions, look at the Manuals on
the R homepage, and particularly under the link 'Contributed.' You should
spend a few hours or even days studying some of the shorter articles in the
Contributed section of the CRAN archive. In particular, you need to know
basic things like starting up R, simple arithmetic operations, and quitting
R. It is possible to skip this step and to learn R as you read this book, but
in that case you have to be prepared to look up the online help available
with R. For readers who want to start using the present book immediately,
we provide a very basic introduction to R in Appendix B; the reader should
work through this material before reading further.

1.2 How to read this book

We recommend that the book be read with an R session open on your computer and the code accompanying this book be kept open as an R file. The
code used in this book is available from the homepage:

  http://www.purl.oclc.org/NET/vasishth/VB/

You will get the most out of this book if you run the code as you read along,
pausing to experiment with the code (changing parameters, asking yourself:
"what would happen if I changed this setting?", etc.). Passive reading of the
textbook will probably not yield much. Do not hesitate to re-read chapters!
This material is best digested by revisiting it several times. The book chapters
are intended to be read in order, since the later chapters build on concepts
introduced in earlier parts of the book.

The accompanying website for the book contains (among other things):

1. A blog for asking questions that the authors or other readers can answer,
   or for submitting comments (corrections, suggestions, etc.) regarding the
   book that other readers can also benefit from.
2. Links to useful websites and other material relating to statistics.
3. Additional material that contains a more advanced treatment of some of
   the issues discussed in the textbook.

1.3 Some Simple Commands in R

We begin with a short session that aims to familiarize you with R and very
basic interaction with data.

Let's assume for argument's sake that we have the grades of eleven students
in a final examination for a statistics course. Both the instructor and the
students are probably interested in finding out at least the maximum and

minimum scores. But hidden in these scores is much more information about the students' grades.

Assuming a maximum possible score of 100, let's first start up R and input the scores (which we just made up).

```
> scores <- c(99, 97, 72, 56, 88, 80, 74, 95, 66,
        57, 89)
```

```
[1] 99 97 72 56 88 80 74 95 66 57 89
```

(When you type the above on the R command line you will not see R echo the contents of scores. You can type scores on the command line to print out the contents. Another way to print out the content of a command is to wrap it in parentheses: (scores). We are able to print out the contents of scores above without using either of these methods because we are using Sweave with LATEX, and Sweave allows the user to print when needed. See the Sweave homepage, http://www.stat.uni-muenchen.de/~leisch/Sweave/, for details)

Now we ask the following questions using R: (a) what's the maximum score? (b) what's the minimum?

```
> max(scores)
```

```
[1] 99
```

```
> min(scores)
```

```
[1] 56
```

We could stop here. But there is much more information in this simple dataset, and it tells us a great deal more about the students than the maximum and minimum grades.

The first thing we can ask is: what is the average or mean score? For any collection of numbers, their MEAN is the sum of the numbers divided by the length of the vector:

$$\bar{x} = \frac{x_1 + x_2 + \cdots + x_n}{n} = \frac{1}{n}\sum_{i=1}^{n} x_i \tag{1.1}$$

The notation $\sum_{i=1}^{n}$ is simply an abbreviation for the statement that the numbers going from x_1 to x_n should be added up.

The mean tells you something interesting about that collection of students: if they had all scored high marks, say in the 90's, the mean would be high, and if not then it would be relatively low. The mean gives you one number that summarizes the data succinctly. We can ask R to compute the mean as follows:

```
> mean(scores)
```

[1] 79.36364

Another such summary number is called the VARIANCE. It tells you how far away the individual scores are from the mean score on average, and it's defined as follows:

$$\text{variance} = \frac{(x_1 - \bar{x})^2 + (x_2 - \bar{x})^2 + \cdots + (x_n - \bar{x})^2}{n-1} = \frac{1}{n-1}\sum_{i=1}^{n}(x_i - \bar{x})^2 \qquad (1.2)$$

The variance formula gives you a single number that tells you how 'spread out' the scores are with respect to the mean. The smaller the spread, the smaller the variance. So let's have R calculate the variance for us:

```
> var(scores)
```

[1] 241.6545

Notice that the number is much larger than the maximum possible score of 100; this is not surprising because we are squaring the differences of each score from the mean when we compute variance. It's natural to ask what the variance is in the same scale as the scores themselves, and to achieve this we can simply take the square root of the variance. That's called the STANDARD DEVIATION (SD), and it's defined like this:

$$s = \sqrt{\frac{\sum_{i=1}^{n}(x_i - \bar{x})^2}{n-1}} \qquad (1.3)$$

Here's how to compute it in R; you can easily verify that it is indeed the square root of the variance:

```
> sd(scores)
```

[1] 15.54524

```
> sqrt(var(scores))
```

[1] 15.54524

At this point you are likely to have at least one question about the definition of variance (1.2). *Why do we divide by $n-1$ and not n?* One answer to this question is that the sum of deviations from the mean is always zero, so if we know $n-1$ of the deviations, the last deviation is predictable. The mean is an average of n unrelated numbers and that's why the formula for mean sums up all the numbers and divides by n. But s is an average of $n-1$ unrelated numbers. The unrelated numbers that give us the mean and standard deviation are also called the DEGREES OF FREEDOM (see Walker, 1940 on a more technical discussion relating to degrees of freedom).

Let us convince ourselves of the observation above that the sum of the deviations from the mean always equals zero. To see this, let's take a look at the definition of mean, and do some simple rearranging.

1. First, look at the definition of mean:

$$\bar{x} = \frac{x_1 + x_2 + \cdots + x_n}{n} \tag{1.4}$$

2. Now we multiply both sides with n:

$$n\bar{x} = x_1 + x_2 + \cdots + x_n \tag{1.5}$$

3. If we now subtract $n\bar{x}$ from both sides:

$$n\bar{x} - n\bar{x} = x_1 + x_2 + \cdots + x_n - n\bar{x} \tag{1.6}$$

4. we get

$$0 = x_1 - \bar{x} + x_2 - \bar{x} + \cdots + x_n - \bar{x} \tag{1.7}$$

The above fact implies that if we know the mean of a collection of numbers, and all but one of the numbers in the collection, the last one is predictable. In equation (1.7) above, we can find the value of ('solve for') any one x_i if we know the values of all the other x's.

Thus, when we calculate variance or standard deviation, we are calculating the average deviation of $n - 1$ unknown numbers from the mean, hence it makes sense to divide by $n - 1$ and not n as we do with mean. We return to this issue again on page 61.

There are other numbers that can tell us about the center-point of the scores, and their spread. One measure is the MEDIAN. This is the midpoint of a sorted (increasing order) list of a distribution. For example, the list 1 2 3 4 5 has median 3. In the list 1 2 3 4 5 6 the median is the mean of the two center observations. In our running example:

```
> median(scores)
```

```
[1] 80
```

The QUARTILES Q_1 and Q_3 are measures of spread around the median. They are the median of the observations below (Q_1) and above (Q_3) the 'grand' median. We can also talk about spread about the median in terms of the INTERQUARTILE RANGE (IQR): $Q_3 - Q_1$. It is fairly common to summarize a collection of numbers in terms of the FIVE-NUMBER SUMMARY: Min Q_1 Median Q_3 Max.

The R commands for these are shown below; here you also see that the command **summary** gives you several of the measures of spread and central tendency we have just learnt.

```
> quantile(scores, 0.25)
```

```
25%
 69
```

```
> IQR(scores)
```

```
[1] 23
```

```
> fivenum(scores)
```

```
[1] 56 69 80 92 99
```

```
> summary(scores)
```

```
  Min. 1st Qu.  Median    Mean 3rd Qu.    Max.
 56.00   69.00   80.00   79.36   92.00   99.00
```

1.4 Graphical Summaries

Apart from calculating summary numbers that tell us about the center and spread of a collection of numbers, we can also get a graphical overview of these measures. A very informative plot is called the boxplot (Figure 1.1): it essentially shows the five-number summary. The box in the middle has a line going through it, that's the median. The lower and upper ends of the box are Q1 and Q3 respectively, and the two 'whiskers' at either end of the box extend to the minimum and maximum values.

```
> boxplot(scores)
```

Another very informative graph is called the HISTOGRAM. What it shows is the number of scores that occur within particular ranges. In our current example, the number of scores in the range 50-60 is 2, 60-70 has 1, and so on. The `hist` function can plot it for us; see Figure 1.2.

```
> hist(scores)
```

Collections of scores such as our running example can be described graphically quite comprehensively, and/or with a combination of measures that summarize central tendency and spread: the mean, variance, standard deviation, median, quartiles, etc. In the coming chapters we use these concepts repeatedly as we build the theory of hypothesis testing from the ground up. But first we have to acquire a very basic understanding of probability theory (chapter 2).

At this point it may be a good idea to work through Baron and Li's tutorial on basic competence in R (there are many other good tutorials available on the CRAN website; feel free to look at several of them). The tutorial is available in the 'Contributed' section of the CRAN website. Here is the link: http://www.psych.upenn.edu/~baron/rpsych/rpsych.html.

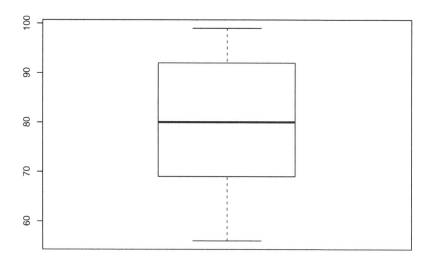

Fig. 1.1 A boxplot of the *scores* dataset.

Once you have read that tutorial, you should install the library vb; this library is available from the book's homepage and contains all the custom functions used in the coming chapters. Finally, you should also install the following packages within R:

1. faraway
2. lme4 (For installing this library on Mac OS X, you first need to install some other software; see http://r.research.att.com/tools/.)
3. lattice
4. UsingR
5. languageR

Instructions on how to install packages are provided on this book's website; alternatively, you can look up how to use the function install.packages in R.

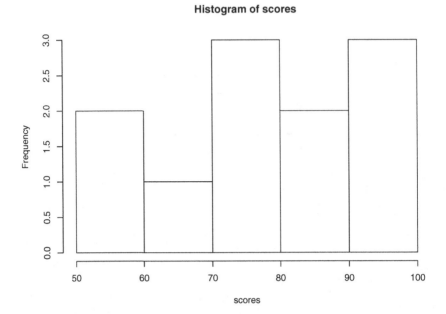

Fig. 1.2 A histogram of the *scores* dataset.

Chapter 2
Randomness and Probability

Suppose that, for some reason, we want to know how many times a second-language learner makes errors in a writing task; to be more specific, let's assume we will only count verb inflection errors. The dependent variable (here, the number of inflection errors) is random in the sense that we don't know in advance exactly what its value will be each time we assign a writing task to our subject. The starting point for us is the question: What's the pattern of variability (assuming there is any) in the dependent variable?

The key idea for inferential statistics is as follows: If we know what a 'random' distribution looks like, we can tell random variation from non-random variation.

As we will see, many random phenomena have the following property: while they are unpredictable in specific individual cases, they follow predictable laws in the aggregate. Once we learn to identify this 'pattern of chance,' we can confidently distinguish it from patterns that are not due to random phenomena.

In this chapter and the next, we are going to pursue this key idea in detail. Our goal here is to look at distribution patterns in random variation (and to learn some R on the side). Before we get to this goal, we need know a little bit about probability theory, so let's look at that first.

2.1 Elementary Probability Theory

2.1.1 The Sum and Product Rules

We will first go over two very basic facts from probability theory: the sum and product rules. Amazingly, these are the only two facts we need for the entire book. The basic idea is that, just as in propositional logic we build up complex propositions from basic ones and compute the truth value of

S. Vasishth, M. Broe, *The Foundations of Statistics: A Simulation-based Approach*, DOI 10.1007/978-3-642-16313-5_2,
© Springer-Verlag Berlin Heidelberg 2011

the whole from the values of the parts, so in probability theory we build up complex *events* from simple ones and compute the probability of the whole event from the probabilities of the parts, using the sum and product rules. Keep this compositional property in mind as you work through this chapter.

We are going to present these ideas completely informally. There are very good books that cover more detail; in particular we would recommend *Introduction to Probability* by Charles M. Grinstead and J. Laurie Snell. The book is available online from the website

http://www.dartmouth.edu/~chance/.

For the present book you do not need to know anything more than what we discuss here. However, a time will come when you will want to read up more on probability theory; the above-mentioned book would be a useful starting point.

Consider the toss of a fair coin, which has two sides, H(eads) and T(ails). Suppose we toss the coin once. What is the probability of an H, or a T? You might say, 0.5, but why do you say that? You are positing a theoretical value based on your prior expectations or beliefs about that coin. (We leave aside the possibility that the coin lands on its side.) We will represent this prior expectation by saying that $P(H) = P(T) = \frac{1}{2}$.

Now consider what all the logically possible outcomes are: an H or a T. What's the probability that either one or the other of these will occur when we toss a coin? Of course, you'd say, 1; we're one hundred percent certain it's going to be an H or a T. We can express this intuition as an equation, as the sum of two mutually exclusive events:

$$P(H) + P(T) = 1 \tag{2.1}$$

There are two things to note here. One is that the two events are *mutually exclusive*; you can't have an H and a T in any one coin toss. The second is that these two events exhaust all the logical possibilities in this example. The important thing to note is that **the probability of mutually exclusive events occurring is the sum of the probabilities of each of the events**. This is called the SUM RULE.

To understand this idea better, think of a fair six-sided die. The probability of each side s is $\frac{1}{6}$. If you toss the die once, what is the probability of getting an odd number? The event 'getting an odd number' can be broken down into the mutually exclusive events 'getting a 1, or a 3, or a 5' and so the answer is $\frac{1}{6} + \frac{1}{6} + \frac{1}{6} = \frac{1}{2}$.

Suppose now that we have not one but *two* fair coins and we toss each one once. What are the logical possibilities now? In other words, what sequences of heads and tails are possible? You'll agree that the answer is: HH, HT, TH, TT, and also that *all of these are equiprobable*. In other words: $P(HH) = P(HT) = P(TH) = P(TT)$. There are four possible EVENTS and each is equally likely. This implies that the probability of each of these is $P(HH) = P(HT) = P(TH) = P(TT) = \frac{1}{4}$. If you see this intuitively, you also understand intuitively the concept of PROBABILITY MASS. As the word 'mass' suggests, we have redistributed the total 'weight' (1) equally over all the logically possible outcomes (there are 4 of them).

Now consider this: the probability of any one coin landing an H is $\frac{1}{2}$, and of landing a T is also $\frac{1}{2}$. Suppose we toss the two coins one after another as discussed above. What is the probability of getting an H with the first coin followed by a T in the second coin? We could look back to the previous paragraph and decide the answer is $\frac{1}{4}$. But probability theory has a rule that gives you a way of calculating the probability of this event:

$$P(H) \times P(T) = \frac{1}{2} \times \frac{1}{2} = \frac{1}{4} \tag{2.2}$$

In this situation, an H in the first coin and an H or T in the second are completely independent events—one event cannot influence the other's outcome. This is the PRODUCT RULE, which says that **when two or more events are independent, the probability of both of them occurring is the product of their individual probabilities.**

And that's all we need to know for this book. At this point, you may want to try solving a simple probability problem: Suppose we toss three coins; what are the probabilities of getting 0, 1, 2, and 3 heads?

2.1.2 Stones and Rain: A Variant on the Coin-toss Problem

Having mastered the two facts we need from probability theory, we now begin our study of randomness and uncertainty, using SIMULATIONS. When you repeatedly carry out random simulations, you will usually get a different answer each time. Repeatedly running such a simulation (performing many REPLICATES) gives you a good intuitive sense of how patterns change or remain stable on each run, and the extent to which they change. We encourage you to run simulations more than once, to see these effects for yourself. We also encourage you to change the values given to the functions (altering the input probabilities, changing the sample size, etc.). This is the great advantage of using a language like R to explore statistical concepts.

Because the coin example is so tired and over-used, we will use a different example of a random process (due originally to Polya, 1954, 56) for purposes of our discussion of probability. Suppose we have two identical stones (labeled L(eft) and R(ight)), and some rain falling on them. We will now create an artificial world in R and 'observe' the raindrops falling randomly on the stones. We can simulate this quite easily in R using the built-in random generator function rbinom(), which takes three arguments:

```
> rbinom(1, 1, 0.5)
```

[1] 1

The above command says that, in a world in which the probability of a R-stone hit is 0.5 (a reasonable assumption given the stones are the same size), generate a raindrop on one pair of stones (the second argument), and do this once (the first argument). Return how many times we successfully get such a hit. So a return value of 1 means the drop landed on the Right stone, 0 means it didn't.

We can repeat this experiment 10 times as follows:

```
> rbinom(10, 1, 0.5)
```

```
[1] 1 1 1 1 1 1 0 1 1 1
```

Notice that different repetitions give different results (it is a random process after all):

```
> rbinom(10, 1, 0.5)
```

```
[1] 0 0 0 1 1 0 0 1 0 0
```

Intuitively, if the probability of a success is 0.5, then this should be reflected in the proportion of successes we see in our observations:

```
> sum(rbinom(10, 1, 0.5))/10
```

```
[1] 0.3
```

And if we increase the number of replicates, this proportion gets closer and closer to the actual probability underlying the process:

```
> sum(rbinom(100, 1, 0.5))/100
```

```
[1] 0.46
```

```
> sum(rbinom(1000, 1, 0.5))/1000
```

```
[1] 0.478
```

```
> sum(rbinom(1e+06, 1, 0.5))/1e+06
```

```
[1] 0.500579
```

In our next experiment, we will observe a section of pavement consisting of 40 pairs of stones, and record the total number of successes (Right-stone hits) for that section (we just keep track of where the first drop falls in each case). Note that we could just as well observe one pair of stones 40 times: the important point is that our single experiment now consists of a group of 40 distinct and independent TRIALS.

```
> rbinom(1, 40, 0.5)
```

```
[1] 24
```

Note that there are many different OUTCOMES of the experiment that could have given rise to the hit-count we just generated. R is not telling us the pattern of Right-stone hits on the section of pavement (or the particular sequence of hits and misses on one pair of stones): it is simply reporting the count, which is usually all we are interested in. Many different patterns have the same count. (We will return to the structure of the individual outcomes later).

Intuitively, since the probability is 0.5, we would expect about half (i.e., 20) stones to register successful hits in any one experiment. But notice too that there are actually 41 *possible* counts we could see: it is possible (though we think it unlikely) that in our observation of 40 stones, we see just 3, or as many as 39, successful hits. We can get a feel for this if we repeat this experiment a number of times:

```
> rbinom(10, 40, 0.5)
```

```
[1] 19 23 15 20 22 19 22 25 22 20
```

The number of successes moves around across a range of numbers (it is a random process after all). You may be surprised to see how seldom we observe 20 hits (in a short time we will be able to compute the exact probability of this occurring). But also notice that it tends not to move across the entire possible range (0–40). In fact, it moves across a range of numbers that are rather close together, and 'balanced' around a 'central' number. The process is random in the details, but has a predictable pattern in the mass. Let's do more random simulations to explore this further.

First we create a vector to hold the results of 1000 experiments:

```
> results <- rbinom(1000, 40, 0.5)
```

Let us look at a histogram of these results:

```
> hist(results, breaks = 0:40)
```

Figure 2.1 shows the result of the above command. The central number and the balance of the distribution become apparent: the most frequently occurring value in this list is (about) 20, and it is centered and balanced in the sense that the frequency of values above and below it fall off symmetrically.

The extent to which this pattern is apparent depends on the number of times we repeat (replicate) the experiment. Let's replicate this experiment i times, where i =15,25,100,1000. Each time we will plot the histogram of the results.

The code for plotting the distributions is shown below:

```
> multiplot(2, 2)
> p <- 0.5
```

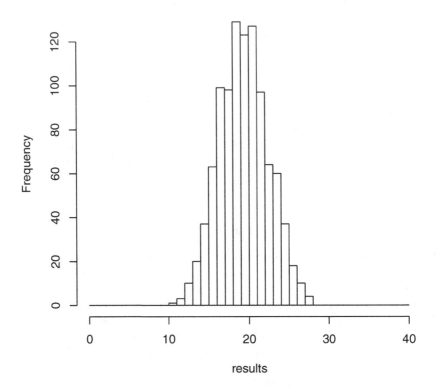

Fig. 2.1 The frequency histogram of the result of observing 40 drops 1000 times.

```
> drops <- 40
> replicates <- c(15, 25, 100, 1000)
> for (i in replicates) {
        results <- rbinom(i, drops, p)
        title <- paste(c("No. Obs.", i, sep = " "))
        hist(results, breaks = 0:40, ylab = "Frequency",
            xlab = "No. of R-stone hits", main = title)
    }
```

The resulting plot is shown in Figure 2.2. Let's take a moment to understand what this code does before we go any further.

1. We are going to plot four different histograms, corresponding to the four values $i = 15, 25, 100, 1000$. We need to instruct R to plot a 2×2 plot. The code for this would be:

```
> par(mfrow = c(2, 2), pty = "s")
```

This command seems rather obscure, so it might be a good idea to write an easier-to-remember function that does the same thing (we will use this function in future):

```
> multiplot <- function(row, col) {
      par(mfrow = c(row, col), pty = "s")
  }
```

2. The next few lines are just fixed values for the simulations; p is the probability (0.5) of a R-stone hit; drops is the number of drops in each experiment (in our example, 40); and replicates is the number of times we repeat the experiment.

```
> p <- 0.5
> drops <- 40
> replicates <- c(15, 25, 100, 1000)
```

3. A *for*-loop then begins by setting the variable i to the first item in the vector replicates (15), and then runs i experiments involving 40 raindrops each, returning the number of Right-stone hits in each of the i runs. Then the hist function is used to plot the distribution of these i R-stone hits; this distribution is stored in the vector called results. Then the loop begins again, but this time with the second value in the observations vector (25) as the value for i; then the above procedure is repeated. Eventually, the final value in the observations vector (1000) is used as the value for i.

The stabilization about a central value that you see in Figure 2.2 is typical of random phenomena. The central value here is 20. In our coin-tossing example, recall that we said that 'intuitively, if the probability of a success is 0.5, then this should be reflected in the *proportion* of successes we see in our observations.' And we saw that, the more replicates we performed, the more closely this proportion approached a limiting number, the probability. In exactly the same way, the proportion of 20 successes, the proportion of 21, of 22 etc., in our current simulation, all approach fixed numbers. We can get a rough sense of what these are by plotting the RELATIVE FREQUENCY HISTOGRAM of our previous results (Figure 2.3):

```
> hist(results, breaks = 0:40, freq = FALSE)
```

By inspection it appears that the relative frequency of the most common result is a little over 0.12 (recall how seldom the exact count 20 occurred in our initial simulations—we now see it occurs a little over 10% of the time.) The relative frequencies fall away in a symmetrical pattern and seem to become effectively zero for counts below 10 and above 30. Where does this pattern come from? Can we characterize it exactly? A common definition

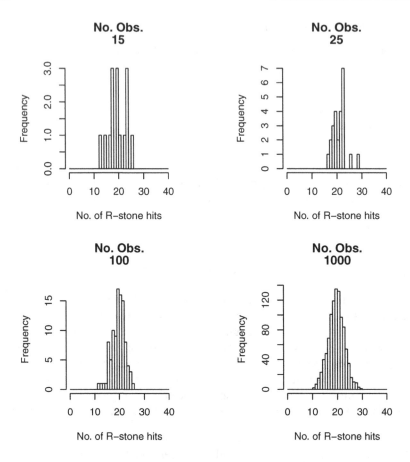

Fig. 2.2 The frequency of Right-stone hits as the number of replicates increases from 15 to 1000. Note that as the number of replicates increases, the most frequently occurring number of Right-stone hits is in the range of 20—exactly half the total number of drops observed each time.

of probability is the limiting final value of relative frequency. Can we compute the exact probability of each of these counts, without doing an infinite number of simulations? It turns out we can, using our elementary sum and product rules, but the computation is intensive for large numbers (which is why we have R). In order to do this 'by hand' let's focus on the case where the number of trials is just 4.

We have seen that when we simulate a random process, we cannot predict what any particular outcome will be: LRRL, RLLL, RRLL, etc. But we have also seen that not all of these outcomes are equal. If we assign each outcome to its 'success count' (2, 1, 2 etc.), we saw that over many repetitions, some of these counts turn up more frequently than others. Why is this, and what

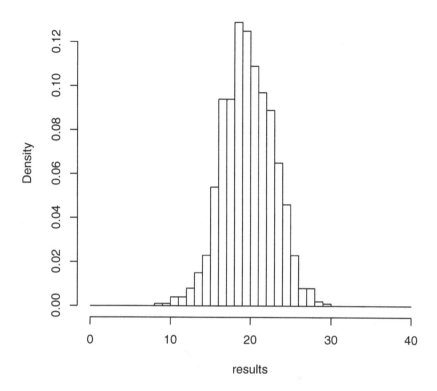

Fig. 2.3 The relative frequency histogram of the result of observing 40 drops 1000 times.

are the exact proportions? There are five possible such counts: 0, 1, 2, 3, 4. What is the exact probability of each?

Let's focus on one particular count. What is the probability of the event E = 'count is 3'? This is actually a complex event, made up of both sums and products of primitive probabilities. Suppose that in one outcome or event E_1 we see RRRL (in that order). That is:

$$E_1 = (RRRL) \tag{2.3}$$

What's the probability of this happening? Well, we know that $P(L) = P(R) = \frac{1}{2}$, and we know the multiplication rule for independent events.

$$P(E_1) = P(R) \times P(R) \times P(R) \times P(L) = \frac{1}{16} \tag{2.4}$$

But there are four distinct outcomes that yield three Rights-stone hits, call them E_1, E_2, E_3, E_4: $E_1 = $ RRRL, $E_2 = $ RRLR, $E_3 = $ RLRR, $E_4 = $ LRRR. So we have a complex event E, made up of four mutually exclusive possibilities:

$$E = E_1 \text{ or } E_2 \text{ or } E_3 \text{ or } E_4 \tag{2.5}$$

which means we can use the summation rule:

$$P(E) = P(E_1) + P(E_2) + P(E_3) + P(E_4) = \frac{1}{4} \tag{2.6}$$

You can already see that figuring out the answer is going to be a pretty tedious business. Let's think of a better way to work this out. Consider the visualization in Figure 2.4 of the PROBABILITY SPACE when we carry out four trials.

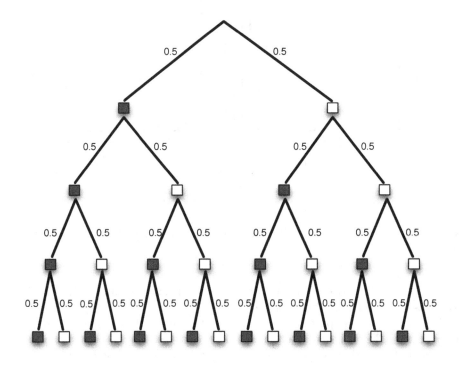

Fig. 2.4 The probability space for four raindrops falling on Left (black) and Right (white) stones.

Figure 2.4 helps us work out the relevant answers. The top of the tree represents the initial state of the probability space, when no raindrop has been observed. When we observe a raindrop once, we can get either a Left-stone hit or a Right-stone hit, and these mutually exclusive events are represented by the two boxes (black for a Left-stone hit and white for a Right-stone hit). Each is an equiprobable event. After one of these possible events, another trial will yield either a Left-stone hit or a Right-stone hit, and so on for the other two trials. So if we go from the top to the bottom of this tree, following each possible path, we have all the logically possible outcomes of Left- and Right-stone hits in this four-trial example. Thus we get: probability of zero Right-stone hits: 0.5^4; probability of only one Right-stone hit: 4×0.5^4; probability of exactly two Right-stone hits: 6×0.5^4; probability of three Right-stone hits: 4×0.5^4; and probability of four Right-stone hits: 0.5^4.

If we multiply the probabilities along each path of the tree and then add them up, they will sum to 1, since these exhaust the possibilities. This visualization method generalizes to any number of trials, and to experiments involving more than two possible outcomes (e.g., the toss of a die).

2.2 The Binomial Distribution

An essential part of this computation then, is knowing *how many* outcomes yield the same count: how many ways are there to have three R's and one L, say. It turns out there are 4 outcomes of count 3 (as we worked out by hand). All such outcomes have the same probability, $\frac{1}{16}$, so the probability of this count will be $\frac{1}{16}$ multiplied by a factor of 4. This factor is well known from combinatorial theory. We want the number of ways we can arrange 3 R's in 4 positions. This is $\binom{4}{3}$ (read "four choose three"), known as the BINOMIAL COEFFICIENT (its formula can be found in any text on discrete math or combinatorial theory such as Rosen, 2006). It is available as a built-in function in R:

```
> choose(4, 3)
```

```
[1] 4
```

The number of ways of arranging $0 \ldots 4$ R's in 4 positions is:

```
> choose(4, 0:4)
```

```
[1] 1 4 6 4 1
```

(The 'positions' here are the physical arrangement of trial stones on the pavement, or the temporal sequence of trials using one pair of stones). Returning to our larger example, the number of possible 13 Right-stone hits in 40 trials is:

```
> choose(40, 13)
```

```
[1] 12033222880
```

And we can display the number of ways of arranging 0:40 R's in 40 positions
(Figure 2.5):

```
> num.outcomes <- choose(40, 0:40)
> plot(num.outcomes)
```

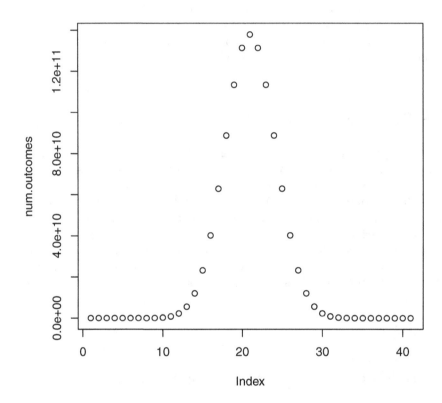

Fig. 2.5 The binomial coefficient of 0:40 Right-stone hits.

To get the exact probabilities of the counts, we just need to multiply the
number of outcomes corresponding to each count by each outcome's individual probability. Assuming equiprobable primitive events, which simplifies
things a little, this will be, say, $(0.5^{19} \times 0.5^{21})$ for the count 19, $(0.5^{20} \times 0.5^{20})$
for the count 20, etc., which are all equal to each other, and all equal to 0.5^{40}.

```
> p <- (0.5^40)
```

```
[1] 9.094947e-13
```

Note the minuscule probability of any particular outcome! This is the probability of getting the *exact* sequence LRLLRRLRLR...RLR of length 40, for example. Finally we combine this number with the binomial coefficients to produce the exact probabilities for each count (Figure 2.6).

```
> binom.distr.40 <- choose(40, 0:40) * p
> plot(binom.distr.40)
```

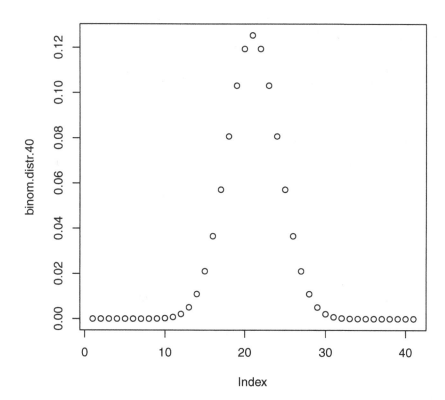

Fig. 2.6 The binomial distribution for $n = 40$.

Since these are all of the mutually exclusive possibilities, the sum of their probabilities should equal 1.

```
> sum(binom.distr.40)
```

```
[1] 1
```

And now we can compute the *exact* probability of observing exactly 20 hits:

```
> choose(40, 20) * p
```

```
[1] 0.1253707
```

To summarize, the BINOMIAL THEOREM allows us to compute the probability of k Right-stone hits (successes) when we make n observations (trials), when the probability of a Right-stone hit (success) is p:

$$\binom{n}{k} \times p^k (1-p)^{n-k} \tag{2.7}$$

The binomial theorem can be applied whenever there are only two possible primitive outcomes, the fixed, n trials are mutually independent, and the probability p of a 'success' is the same for each trial.

Recall the pattern that emerged as we simulated raindrops: as the number of replicates (the number of times we observe 40-drop sequences) increases from 15 to 1000, the *relative frequencies* of R-stone hits settle down to stable values. The *distribution* of R-stone hits has a stable final shape. Just as we express each final *value* in terms of a *theoretical probability*, so we can express the final *shape* in terms of a *theoretical probability distribution* (which we approached empirically using simulation and have now derived mathematically from the primitive probabilities). This shows us how the total probability is distributed among all the possible results of an experiment

The 'central value' of the distribution is in fact the MEAN of the distribution:

```
> mean(results)
```

```
[1] 20.181
```

And as the number of replicates increases, the closer this mean approaches the theoretical limit of 20. The **mean of the sampling distribution** will be explored in great detail as we go forward.

2.3 Balls in a Box

Let's now exercise our new-found knowledge about the binomial distribution in a different scenario, and formally introduce the concepts of sampling theory.

Suppose we have 12,000 balls in a big box, and we *know* that 9000 (i.e., 3/4) are Red, the others White. We say we have a POPULATION of 12,000. Suppose we take a RANDOM SAMPLE of 100 balls from these 12,000. We'd expect to draw about 75 white balls. What's the probability of getting *exactly* 75?

We can simulate this scenario by repeatedly drawing random samples from such a box. For each sample we compute a number, the count of Red balls or 'successes.' A number that describes some aspect of a sample is called a STATISTIC. The particular statistic we are computing here is the SAMPLE COUNT, and if we plot the results we will be able to get an idea of the SAMPLING DISTRIBUTION of this statistic.

The result of the simulation is shown in Figure 2.7, and once again it's worth pausing to examine the code (shown below). In this simulation, and in many others we will encounter, the `for`-loop steps through the replicates, and for each replicate the result is added to a vector, here `sample.counts`. This vector needs to be *initialized* before we can add values to it: this happens prior to the `for`-loop, where we fill the vector with 1000 missing values, symbolized by `NA` in R, using the `rep()` function (short for 'repeat'). We use the same function to fill the box with $9,000$ Reds and $3,000$ Whites. We code Red as 1, and White as 0, purely for convenience, since then we can compute the sample count of Reds simply by summing the sample itself. The function `sample()` is used to extract a sample of the specified size. Finally, we plot the frequency and relative frequency histogram of the sample counts.

```
> box <- c(rep(1, 9000), rep(0, 3000))
> sample.counts <- rep(NA, 1000)
> for (i in 1:1000) {
      a.sample <- sample(box, 100)
      sample.counts[i] <- sum(a.sample)
  }
> multiplot(1, 2)
> hist(sample.counts, breaks = 50:100)
> hist(sample.counts, breaks = 50:100, freq = FALSE)
```

Inspection of the distribution shows that the central and most probable value is indeed close to 75, with a relative frequency of about 0.09. The pattern approximates a binomial distribution. Can we compute the exact probability of obtaining a sample with exactly 75 balls? As we now know, this partly depends on *how many ways* it is possible to draw a sample whose count equals 75 Reds. We need the binomial theorem:

$$\binom{n}{k} \times p^k (1-p)^{n-k} \tag{2.8}$$

Let's first define a function in R to calculate this quickly:

```
> binomial.probability <- function(k, n, p) {
      choose(n, k) * p^k * (1 - p)^(n - k)
  }
```

If we run that function now with $k = 75$, $n = 100$, $p = 3/4$, we find that the probability of getting *exactly* 75 is:

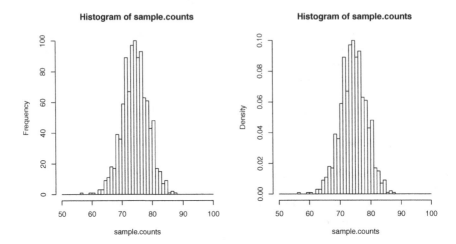

Fig. 2.7 The sampling distribution for the sample count of Red balls in a sample of 100, showing absolute frequency (on the left) and relative frequency (on the right).

```
> binomial.probability(75, 100, 3/4)
```

```
[1] 0.0917997
```

As an aside, note that R provides this function under the name of dbinom:

```
> dbinom(75, 100, 3/4)
```

```
[1] 0.0917997
```

Now we ask an interesting question: suppose the sample was larger, say 1000. Intuitively, by increasing the sample size, we should get a better estimate of what is actually in the box. After all, if the sample size was 12,000, we would *know* the contents of the box, and the larger our sample, the closer we should get to certainty. Would the probability of drawing 3/4 Red balls (750) from this larger sample be higher or lower than the number (0.0918) that we computed for a sample 10 times smaller? Think about this before reading further...

Here is the answer:

```
> binomial.probability(750, 1000, 3/4)
```

```
[1] 0.02912411
```

Thus, as the sample size goes up, the probability of the most likely sample count goes *down*. And this goes for the probabilities of *all* the counts in the larger sample. The reason for this is that a total probability of 1 is being distributed among 101 possibilities in the first case, but among 1001 possibilities in the second. The probability is 'stretched thin' in the larger sample. 750 Reds is still the *most probable* sample count, but in absolute terms, its probability decreases. This leads us to a fundamental principle of statistical inference: we are hardly ever interested in the probability of a *single value*. Instead, we focus on a *range* or INTERVAL of possible values, and compute the probability of being within that range.

We first focus on how to compute the probability of such an interval; we then examine how this probability behaves as we alter the sample size.

To explore this approach, consider an alternative (simpler) scenario where we have 12,000 Red and White balls, and exactly half are Red ($p = 0.5$). We take a sample of 40 balls, and calculate the probability of getting $1 \ldots 40$ Reds:

```
> n <- 40
> p <- 0.5
> probs <- binomial.probability(0:n, n, p)
```

This function simply computes `binomial.probability(0, 40, 0.5)`, `binomial.probability(1, 40, 0.5)`, `binomial.probability(2, 40, 0.5)`, etc., and puts each probability in the vector `probs`. In this and all the code in the rest of this chapter, n represents the size of the sample we draw.

The variable `probs` now contains a vector of probabilities. (Note that the probability of the count 20 is indexed as `probs[21]`. This is because the probability corresponding to 0 Reds is in `probs[1]`. There is no such thing as `probs[0]`. This kind of offset is a commonly encountered in computer coding). Here are the central values of this vector.

```
> probs[20:22]
```

```
[1] 0.1194007 0.1253707 0.1194007
```

The probability of getting exactly 20 Red balls is 0.1254. As we have come to expect, this probability is quite low. Recall too that the value 20 is the mean of the sampling distribution. Let's investigate the probability of an interval around this value.

What's the probability of getting—not exactly 20—but 19 or 20 or 21 Red balls in a sample of 40? This is an interval centered on 20, with a MARGIN OF ERROR ± 1. Since these are three mutually exclusive alternatives, we can use the sum rule:

```
> sum(probs[20:22])
```

```
[1] 0.364172
```

And for margin of error ± 2:

```
> sum(probs[19:23])
```

[1] 0.5704095

Let's go ahead and compute the probabilities for *all* the margins of error.

```
> mean.index <- 21
> intervals <- rep(NA, 20)
> for (i in 1:20) {
        indices <- seq(mean.index - i, mean.index +
            i, by = 1)
        intervals[i] <- sum(probs[indices])
  }
> conf.intervals <- data.frame(margin = rep(1:20),
        probability = intervals)
> head(conf.intervals)
```

```
  margin probability
1      1   0.3641720
2      2   0.5704095
3      3   0.7318127
4      4   0.8461401
5      5   0.9193095
6      6   0.9615227
```

The important point here is that when we increase the margin of error to be ±6 around the precise expected mean number of Red balls (20), the probability of *a value within this interval* occurring is approximately 0.95. Let's visualize this (Figure 2.8).

```
> main.title <- "Sample size 40"
> plot(conf.intervals$margin, conf.intervals$probability,
        type = "b", xlab = "Margins", ylab = "Probability",
        main = main.title)
> segments(0, 0.95, 5.7, 0.95)
> segments(5.7, 0, 5.7, 0.95)
```

The straight lines in Figure 2.8 mark the margin of error (about 6), which corresponds to 0.95 probability. What this means is, as we take repeated samples from the box, the sample count will be *exactly* 20—the exact mean of the distribution—only about 12% of the time, but within 6 of the mean about 95% of the time. Notice too how quickly the curve rises to 0.95. This is due to the simple but important fact that most of the probability is clustered around the mean in the binomial distribution. So a relatively small interval around the mean captures most of the probability.

It is now time to take one of the most important conceptual steps in statistical inference: a kind of inversion of perspective. *If the sample count is within 6 of the mean 95% of the time, then 95% of the time the mean is within 6 of the sample count.* We can use this fact to infer the mean of the

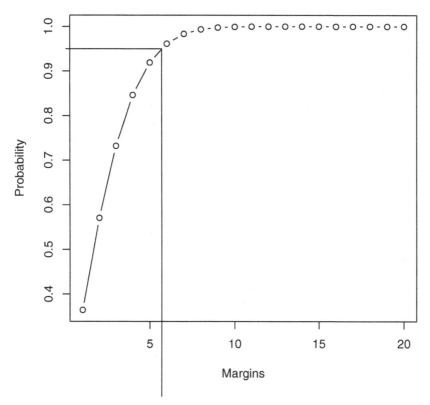

Fig. 2.8 The probability of getting $20 \pm m$ Red balls in a random sample of 40, where m is the margin of error we allow (ranging from 1 to 20).

distribution from a single sample (however, see page 57 for a subtle point relating to this inversion). We cannot know *exactly* what it is, but we can express a 95% LEVEL OF CONFIDENCE that it is in an interval within 6 of the sample count. To take a concrete example:

```
> box = c(rep(1, 6000), rep(0, 6000))
> sum(sample(box, 40))
```

```
[1] 17
```

We do not know what specific value we will get when we execute this code (random processes are unpredictable in specific cases). But exploiting our knowledge of the pattern of random phenomena in the aggregate, we *know* that the value is within 6 of 20, 95% of the time. You may wish to

execute the above code multiple times to prove this to yourself. Note the
logic: we first construct the CONFIDENCE INTERVAL around the mean of the
sampling distribution—this tells us how large the interval must be to capture
the specified amount of the probablity. We then take a single sample, compute
the SAMPLE STATISTIC (the sample count), and center the interval *on the
sample*. This gives us an interval estimate of the value of the mean of the
distribution.

We have constructed a 95% confidence interval above. The figure 95% is an
arbitrary but conventional standard in statistical inference. We could choose
to construct a 99% interval, and if we did, we would have to increase the
size of the interval to capture 0.99 of the probability. Our confidence that
this interval contains the mean, if centered on any individual sample, would
increase, but the *accuracy* of the interval is lower. This is a necessary tradeoff.
What we would like to achieve is both high confidence *and* high accuracy.
How can we achieve this?

What would happen if the sample size were increased from 40 to 400?
Our expected mean number of Red balls would now be 200. The confidence
intervals are computed in the following code, and graphed in Figure 2.9.

```
> n <- 400
> p <- 0.5
> probs <- binomial.probability(0:n, n, p)
> mean.index <- 201
> intervals <- rep(NA, 200)
> for (i in 1:200) {
        indices <- seq(mean.index - i, mean.index +
            i, by = 1)
        intervals[i] <- sum(probs[indices])
    }
> conf.intervals <- data.frame(margin = rep(1:200),
        probability = intervals)
> head(conf.intervals)

  margin probability
1      1   0.1192112
2      2   0.1973747
3      3   0.2736131
4      4   0.3472354
5      5   0.4176255
6      6   0.4842569

> main.title <- "Sample size 400"
> plot(conf.intervals$margin, conf.intervals$probability,
        type = "b", xlab = "Margins", ylab = "Probability",
        main = main.title)
> segments(-6, 0.95, 19, 0.95)
> segments(19, 0, 19, 0.95)
```

Inspection of Figure 2.9 and the full list of intervals computed in the code show that the 95% margin of error in this case is about ±19.

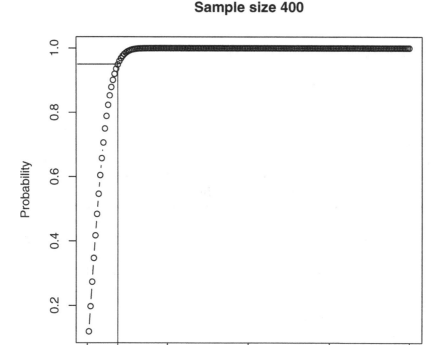

Fig. 2.9 The probability of getting $200 \pm m$ Red balls from a random sample of 400, where m is the margin of error we allow (ranging from 1 to 200).

(In the above examples, we essentially used the same R code twice, with just a few changes. In such situations it makes sense to write a function that can do the same thing, but with different settings—here, different sample sizes. Let's write such a function. The explanation for each computation is shown in the source code at the course website, but try to work it out yourself first).

```
> compute.margins <- function(sample.size, p) {
      probs <- binomial.probability(0:sample.size,
          sample.size, p)
```

```
mean.index <- (sample.size * p) + 1
max.margin <- sample.size * p
intervals <- rep(NA, max.margin)
for (i in 1:max.margin) {
    indices <- seq(mean.index - i, mean.index +
        i, by = 1)
    intervals[i] <- sum(probs[indices])
}
conf.intervals <- data.frame(margin = rep(1:max.margin),
    probability = intervals)
return(conf.intervals)
}
```

What then is the effect of sample size on our 95% confidence interval? What we have established is that as the sample size increases, so does the confidence interval, *in absolute terms*. For sample size 40, the margin of error is ± 6; for a sample of 400, the margin of error is ± 19. The accuracy of our estimate seems to have become less precise! In order to properly compare the margins of error, however, we need to NORMALIZE them so that their range is comparable in both cases (currently, in the 40 sample case the margins range from 1 to 20, and in the 400 sample case they range from 1 to 200). This normalization can be done by converting them to proportions. For example, in the 40 sample case, we simply treat the margin ± 1 (19 and 21) as 19/40 and 21/40 respectively; for the 400 sample case, we treat the margin ± 1 (199 and 201) as 199/400 and 201/400 respectively. When we do this, the result is Figure 2.10. (We first define a function `plot.margins()` to display the margin of error).

```
> plot.margins <- function(sample.size, p, color = "black",
      margin, main, interval = TRUE) {
      probs <- binomial.probability(0:sample.size,
          sample.size, p)
      proportions <- 0:sample.size/sample.size
      plot(proportions, probs, type = "l", col = "black",
          xlab = "Proportions", ylab = "Probability",
          main = main)
      if (interval == TRUE) {
          segments(proportions[(sample.size/2 +
              1) - margin], -0.5, proportions[(sample.size/2 +
              1) - margin], 0.06, col = color,
              lty = 1, lwd = 2)
          segments(proportions[(sample.size/2 +
              1) + margin], -0.5, proportions[(sample.size/2 +
              1) + margin], 0.06, col = color,
              lty = 1, lwd = 2)
```

```
        }
    }
```

Then we plot the margins (Figure 2.10):

```
> multiplot(1, 2)
> main.title.40 <- "Sample size 40"
> main.title.400 <- "Sample size 400"
> plot.margins(40, 0.5, margin = 5, main = main.title)
> plot.margins(400, 0.5, margin = 19, main = main.title)
```

There are a number of important insights to take away from Figure 2.10:

1. As sample size increases from 40 to 400, we get proportionally tighter 95% probability regions: the accuracy of the confidence interval increases.
2. We have effectively employed a new statistic: the SAMPLE PROPORTION. The 40 sample count 21, for example, is replaced by 21/40, the 400 sample count 201, is replaced by 201/400. Figure 2.10 actually shows the sampling distribution of this statistic for two different sample sizes. We explore this further in the next section.
3. The mean of the sampling distribution (of the sample proportion) accurately reflects the proportion of balls in the population. Just as a statistic describes some aspect of a sample, a PARAMETER describes some aspect of a *population*. Thus the mean of the sampling distribution 'points to' the POPULATION PARAMETER.
4. Just as the mean of the sampling distribution is of special significance, so is its standard deviation. Intuitively, the smaller the standard deviation, the tighter the confidence interval, and the more accurate the interval estimate of the mean. In fact, we can quantify this exactly. As will become clear in the coming chapters, the 95% probability region corresponds to approximately 2 times the standard deviation (SD) of the sampling distribution. The smaller the SD, the better the estimate.

In the next section we examine exactly how the SD of the sampling distribution varies with sample size.

Aside: Applying the binomial theorem

As mentioned above, when we want to compute the probability of getting 0 to 20 Right-stone hits when we observe 40 raindrops, we can do this using dbinom:

```
> sums <- rep(NA, 21)
> for (i in 0:20) {
      sums[i + 1] <- dbinom(i, 40, 0.5)
  }
> sum(sums)
```

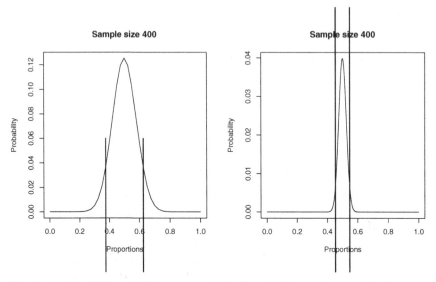

Fig. 2.10 The 95% probability ranges in the 40 and 400 sample case with the margins of error normalized.

[1] 0.5626853

An even easier way to do this in R is:

```
> sum(dbinom(0:20, 40, 0.5))
```

[1] 0.5626853

And yet another way is to say:

```
> pbinom(20, 40, 0.5)
```

[1] 0.5626853

Thus, there is a family of functions for the binomial distribution that we can use to do very useful things:

1. rbinom: the random number generation function
2. dbinom: The probability density function
3. pbinom: The cumulative distribution function (the proportion of values which have a value x or lower)

2.4 Standard Deviation and Sample Size

Earlier we looked at the sampling distribution of the sample count using a 40-drop sequence. Suppose we plot the result of 100 replicates of observing n drops, where n is (a) 4, (b) 40, (c) 400, and (d) 4000 (Figure 2.11).

```
> p <- 0.5
> sample.sizes <- c(4, 40, 400, 4000)
> multiplot(2, 2)
> xlabel <- "Number of R-stone hits"
> for (n in sample.sizes) {
      results <- rbinom(100, n, p)
      maintitle <- paste(n, "drops", sep = " ")
      hist(results, xlim = range(c(0:n)), xlab = xlabel,
          main = maintitle)
  }
```

As we increase the sample size from 4 to 4000 (and observe these n-drop sequences 100 times), the spread, i.e., the standard deviation, appears to decrease. We can see this visually in Figure 2.11. But does it really? Let's plot the standard deviation by sample size, examining all sample sizes from 1 to 100 (Figure 2.12).

```
> sample.sizes <- 1:400
> p <- 0.5
> standard.deviations <- rep(NA, 400)
> for (n in sample.sizes) {
      binomial.distribution.count <- rbinom(100,
          n, p)
      standard.deviations[n] <- sd(binomial.distribution.count)
  }
> plot(sample.sizes, standard.deviations, xlim = c(1,
      400), xlab = "Sample size", ylab = "Standard deviation")
```

To our great surprise, we get increasing values for standard deviation as sample size increases. Look carefully at the scale on the x-axis in each plot. In the first it ranges from 0–4, in the last it ranges from 0–4000. What's going on is that in *absolute* terms the standard deviation of the sample count increases, but as we have seen, raw counts are not directly comparable; we need to relativize the count to the different sample sizes.

If we look at sample proportions rather than sample counts, we'll have *normalized* the various counts so that they are comparable. So, when we take a sample of 40 drops, instead of saying 'the count is 18 Right-stone hits,' we say 'the proportion of Right-stone hits is 18/40.' Let's plot by sample proportion rather than sample count and see what we get. At this juncture you should spend a few minutes trying to modify the previous code in order to

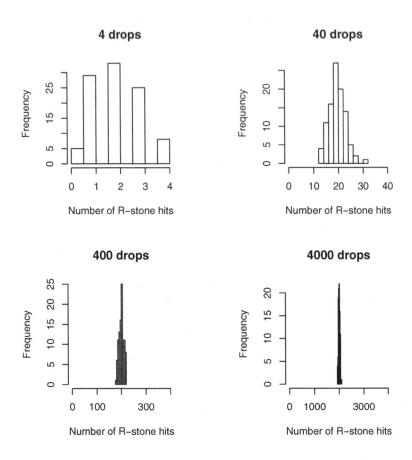

Fig. 2.11 Does increasing the number of drops observed from 4 to 4000 results in a tighter distribution?

plot sample proportions rather than sample counts. The results are shown in (Figure 2.13). Now let's plot the standard deviation of the proportion-based counts (Figure 2.14).

```
> p <- 0.5
> sample.sizes <- c(4, 40, 400, 4000)
> multiplot(2, 2)
> for (n in sample.sizes) {
        results <- rbinom(100, n, p)/n
        maintitle <- paste(n, "drops", sep = " ")
        xlabel <- "Proportion of R-stone hits"
        hist(results, xlim = range(c(0:1)), xlab = xlabel,
             main = maintitle)
    }
```

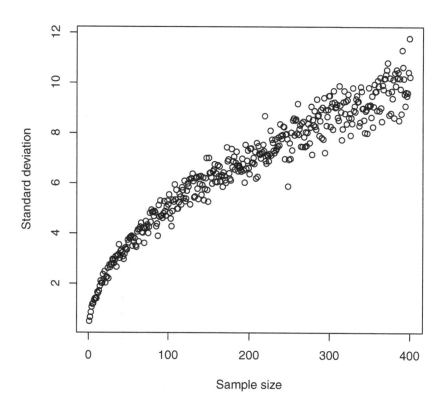

Fig. 2.12 Standard deviation of the sample count increases with sample size.

Plotting the standard deviations of proportions now yields the expected pattern of decreasing standard-deviation values with increasing sample size:

```
> sample.sizes <- 1:400
> p <- 0.5
> standard.deviations <- rep(NA, 400)
> for (n in sample.sizes) {
      binomial.distribution.prop <- rbinom(100,
          n, p)/n
      standard.deviations[n] <- sd(binomial.distribution.prop)
  }
> plot(sample.sizes, standard.deviations, xlim = c(1,
      400), xlab = "Sample size", ylab = "Standard deviation")
```

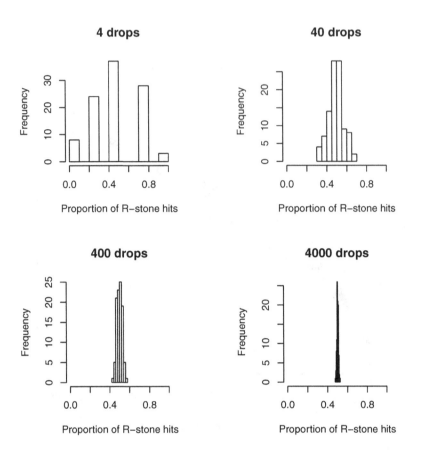

Fig. 2.13 Plot of distribution of sample proportion of Right-stone hits as sample size increases.

Now everything makes sense: look at the scale of the x-axis in these plots. The spread, or standard deviation, decreases as we increase sample size. This is an important insight that we will come back to. We will also revisit this technique of normalizing the scale of comparison in coming chapters.

2.4.1 Another Insight: Mean Minimizes Variance

Recall that variance is defined as follows:

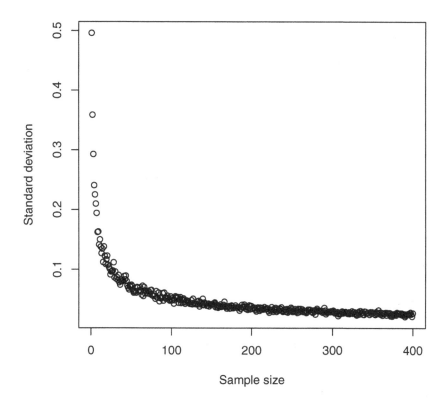

Fig. 2.14 Standard deviation of the sample proportion decreases as sample size increases.

$$s^2 = \frac{(x_1 - \bar{x})^2 + (x_2 - \bar{x})^2 + \cdots + (x_n - \bar{x})^2}{n - 1} = \frac{1}{n - 1} \sum_{i=1}^{n} (x_i - \bar{x})^2 \qquad (2.9)$$

Standard deviation, s, is a measure of spread about the mean, as we just saw. Recall our earlier observation that the sum of deviations from the mean will always equal zero.

A related fact is that the SUM OF SQUARED DEVIATIONS from the mean are smaller than from any other number—the mean is a special number in that sense.

The sum of squared deviations from the mean simply means that, given a vector of scores, we calculate the mean of the vector, and then, subtract each value in the vector from the mean, squaring the result each time; and then we sum up these squared deviations. Consider this simple example; let the vector of scores be a range of values going from 1 to 10:

```
> vector <- 1:10
```

We can compute its mean:

```
> vector.mean <- mean(vector)
```

Then, we can take squared deviations from the mean of each value in the vector, and then sum them up:

```
> sum((vector - vector.mean)^2)
```

```
[1] 82.5
```

We will be using this insight about the sum of squared deviations when we discuss analysis of variance in chapter 5.

Let's quickly convince ourselves that the sum of squared deviations from the mean are smaller than from any other number. First, we generate all the possible squared deviations from the mean and all other possible numbers:

```
> size <- 1
> p <- 0.5
> num.drops <- 4000
> results <- rep(NA, 100)
> for (i in 1:100) {
        results[i] <- sum(rbinom(num.drops, size,
            p))
  }
> mean.results <- mean(results)
> n <- floor(mean.results - 1)
> m <- floor(mean.results + 1)
> xvalues <- c(1:n, mean.results, m:4000)
> all.sq.deviations <- rep(NA, length(xvalues))
> for (i in xvalues) {
        vector.i <- rep(i, 100)
        deviations <- results - vector.i
        sq.deviations <- sum(deviations^2)
        all.sq.deviations[i] <- sq.deviations
  }
```

Next, we plot the sum of squared deviations from the mean against all other numbers (Figure 2.15).

```
> xlabel <- "Potential minimizers of sum of squared deviation"
> plot(xvalues, all.sq.deviations, xlab = xlabel,
        ylab = "Squared Deviation")
> lines(xvalues, all.sq.deviations)
```

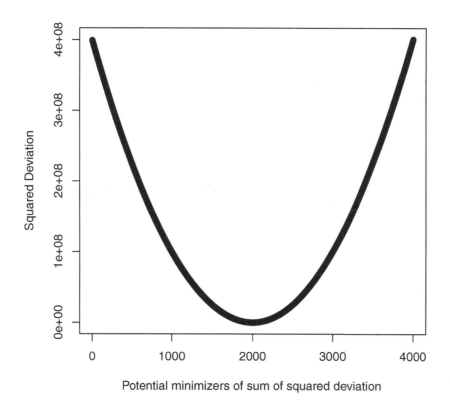

Potential minimizers of sum of squared deviation

Fig. 2.15 The mean minimizes sum of squared deviations: the sum of squared deviations from the mean are smaller than from any other number.

2.5 The Binomial versus the Normal Distribution

Before we proceed further, we would like to introduce a distribution that is very similar to the binomial distribution shown in Figure 2.6. It is defined by this somewhat intimidating-looking function:

$$f(x) = \frac{1}{(\sigma\sqrt{2\pi})} E^{-((x-\mu)^2/2\sigma^2)} \tag{2.10}$$

Given a range of values for x, and specific values for μ, and σ, we could plot the result of applying this function. Let's define this function, plot it, and compare it to the binomial distribution (Figure 2.16).

```
> new.function <- function(x, mu, sigma) {
        1/(sqrt(2 * pi) * sigma) * exp(1)^(-((x -
            mu)^2/(2 * sigma^2)))
    }

> main.title <- "Comparing the binomial and normal distributions"
> plot.margins(40, 0.5, 40, margin = 20, main = main.title,
        interval = FALSE)
> lines(c(1:40)/40, new.function(c(1:40), 20, 3),
        col = "black", lty = 2)
```

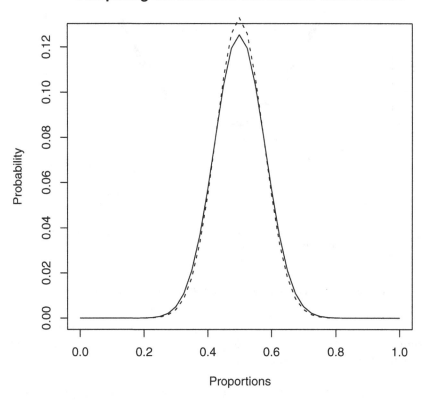

Comparing the binomial and normal distributions

Fig. 2.16 Comparison of the binomial (solid line) and the normal (dashed line) distributions.

This is known as the normal distribution function and has a very similar shape to the binomial distribution. This is not surprising, since the function was originally formulated by De Moivre in the early 1700s as an *approximation* to the binomial. As we have seen, working with the exact discrete binomial involves fantastically large numbers of sums and products, which were impossible to compute before the advent of digital computers. De Moivre's analytic formulation facilitated proofs of the distribution's mathematical properties.

One important difference between the normal and binomial distributions is that the former refers to continuous dependent variables, whereas the latter refers to a discrete binomial variable. An example of a continuous dependent variable would be reaction time or reading time data. In the next chapter we will use the normal distribution to explore the behavior of a new, continuous statistic: the sample mean.

Problems

2.1. Imagine that you have a biased coin, where the probability of obtaining a heads is not 0.5 but 0.1. When the coin is tossed four times, what are the probabilities of obtaining 0, 1, 2, 3, 4 heads?

2.2. Using a probability-space tree like the one shown in Figure 2.4, calculate the probability of obtaining a 6 each time when a die is tossed three times in a row.

2.3. What is the probability of obtaining any of the numbers 2, 4, or 6 if a die is tossed three times in a row?

Chapter 3
The Sampling Distribution of the Sample Mean

We begin by establishing a fundamental fact about any normal distribution: about 95% of the probability lies within 2 SD of the mean. If we integrate the area under these curves, between 2 SD below the mean and 2 SD above the mean, we find the following areas, which correspond to the amount of probability within these bounds:

```
> integrate(function(x) dnorm(x, mean = 0, sd = 1),
      -2, 2)

0.9544997 with absolute error < 1.8e-11

> integrate(function(x) dnorm(x, mean = 0, sd = 4),
      -8, 8)

0.9544997 with absolute error < 1.8e-11
```

We can display this fact graphically (see Figure 3.1):

```
> main.title <- "Area within 2 SD of the mean"
> multiplot(1, 2)
> plot(function(x) dnorm(x, mean = 0, sd = 1),
      xlim = c(-3, 3), main = "SD 1", xlab = "x",
      ylab = "", cex = 2)
> segments(-2, 0, -2, 0.4)
> segments(2, 0, 2, 0.4)
> plot(function(x) dnorm(x, mean = 0, sd = 4),
      xlim = c(-12, 12), main = "SD 4", xlab = "x",
      ylab = "", cex = 2)
> segments(-8, 0, -8, 0.1)
> segments(8, 0, 8, 0.1)
```

Suppose that we have a population of people and that we know the age of each individual; let us assume also that distribution of the ages is approximately normal. Finally, let us also suppose that we know that mean age of the population is 60 and the population SD is 4.

S. Vasishth, M. Broe, *The Foundations of Statistics: A Simulation-based Approach*, DOI 10.1007/978-3-642-16313-5_3,
© Springer-Verlag Berlin Heidelberg 2011

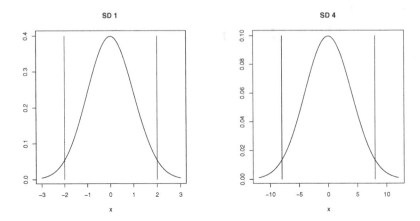

Fig. 3.1 Two normal distributions with SD $= 1$ (left), SD $= 4$ (right). The lines delimit the region 2 SD from the mean in each case.

Now suppose that we repeatedly sample from this population: we take samples of 40, a total of 1000 times; and we calculate the mean $\bar{x}$ each time we take a sample. After taking 1000 samples, we have 1000 means; if we plot the distribution of these means, we have the sampling distribution of the sample mean.

```
> sample.means <- rep(NA, 1000)
> for (i in 1:1000) {
      sample.40 <- rnorm(40, mean = 60, sd = 4)
      sample.means[i] <- mean(sample.40)
  }
```

We can calculate the mean and standard deviation of this sampling distribution:

```
> means40 <- mean(sample.means)
```

```
[1] 59.98692
```

```
> sd40 <- sd(sample.means)
```

```
[1] 0.6156489
```

If we plot this distribution of means, we find that it is roughly normal. We can characterize the distribution of means visually, as done in Figure 3.2 below, or in terms of the mean and standard deviation of the distribution. The mean value in the above simulation is 59.99 and the standard deviation of the distribution of means is 0.6156. Note that if you repeatedly run the above simulation code, these numbers will differ slightly in each run.

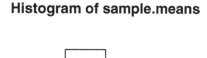

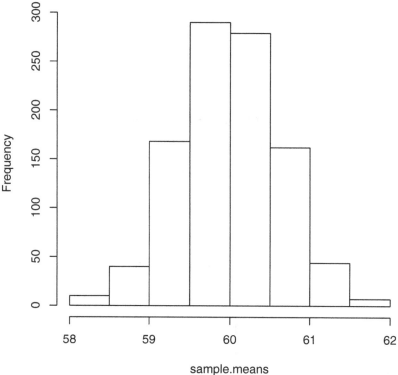

Fig. 3.2 The sampling distribution of the sample mean with 1000 samples of size 40.

```
> hist(sample.means)
```

Consider now the situation where our sample size is 100. Note that the mean and standard deviation of the population ages is the same as above.

```
> sample.means <- rep(NA, 1000)
> for (i in 1:1000) {
      sample.100 <- rnorm(100, mean = 60, sd = 4)
      sample.means[i] <- mean(sample.100)
  }
> means100 <- mean(sample.means)

[1] 59.99521

> sd100 <- sd(sample.means)
```

[1] 0.4065139

In this particular simulation run, the mean of the means is 60 and the standard deviation of the distribution of means is 0.4065. Let's plot the distribution of the means (Figure 3.3).

```
> hist(sample.means)
```

Histogram of sample.means

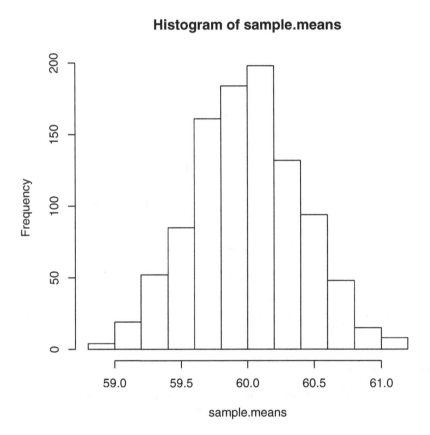

Fig. 3.3 The sampling distribution of the sample mean with samples of size 100.

The above simulations show us several things. First, the standard deviation of the distribution of means gets smaller as we increase sample size. When the sample size is 40, the standard deviation is 0.6156; when it is 100, the standard deviation is 0.4065. Second, as the sample size is increased, the mean of the sample means comes closer and closer to the *population* mean $\mu_{\bar{x}}$. A third point (which is not obvious at the moment) is that there is a lawful relationship between the standard deviation σ of the population and the standard deviation of the *distribution of means*, which we will call $\sigma_{\bar{x}}$.

$$\sigma_{\bar{x}} = \frac{\sigma}{\sqrt{n}} \tag{3.1}$$

Here, n is the sample size. It is possible to derive equation 3.1 from first principles. We do this in Appendix A. For now, simply note the important point that n is in the denominator in this equation, so there is an inverse relationship between the sample size and the standard deviation of the sample means. Let's take this equation on trust for the moment and use it to compute $\sigma_{\bar{x}}$ by using the population standard deviation (which we assume we know). Let's do this for a sample of size 40 and another of size 100:

```
> 4/sqrt(40)
```

```
[1] 0.6324555
```

```
> 4/sqrt(100)
```

```
[1] 0.4
```

The above calculation is consistent with what we just saw: $\sigma_{\bar{x}}$ gets smaller and smaller as we increase sample size.

We have also introduced a notational convention that we will use throughout the book: sample statistics are symbolized by Latin letters ($\bar{x}, s$); population parameters are symbolized by Greek letters (μ, σ).

3.1 The Central Limit Theorem

We've seen in the previous chapter that the distribution of a sample count (and sample proportion) has the shape of a binomial distribution, which is closely approximated by the normal distribution. Now we see that the *sampling distribution of the sample mean* is also normally distributed. In the above example the means were drawn from a population with normally distributed scores. It turns out that the sampling distribution of the sample mean will be normal even if the population is not normally distributed, as long as the sample size is large enough. This is known as the Central Limit Theorem, and it is so important that we will say it twice:

Provided the sample size is large enough, the sampling distribution of the sample mean will be close to normal *irrespective of what the population's distribution looks like.*

Let's check whether this theorem holds by testing it in an extreme case, simulating a population which we *know* is not normally distributed. Let's take our samples from a population (Figure 3.4) whose values are distributed exponentially with the same mean of 60 (the mean of an EXPONENTIAL DISTRIBUTION is the reciprocal of the so-called 'rate' parameter).

```
> sample.100 <- rexp(100, 1/60)
> hist(sample.100)
```

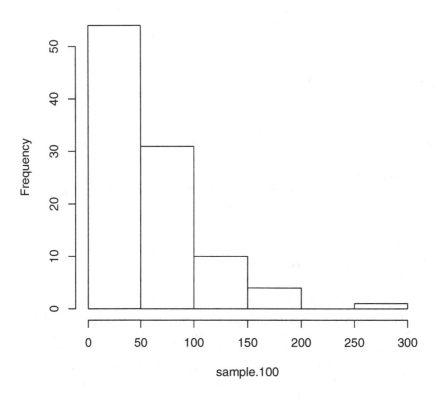

Histogram of sample.100

Fig. 3.4 A sample from exponentially distributed population scores.

Now let us plot the sampling distribution of the sample mean. We take 1000 samples of size 100 each from this exponentially distributed population. As shown in Figure 3.5, the distribution of the means is again normal!

```
> sample.means <- rep(NA, 1000)
> for (i in 1:1000) {
        sample.100 <- rexp(100, 1/60)
```

```
        sample.means[i] <- mean(sample.100)
   }
> hist(sample.means)
```

Recall that the mean of each sample is a 'point summary' of the distribution. Some of these samples will have a mean slightly above the true mean, some slightly below, and the sampling distribution of *these* values is roughly normal. Try altering the sample size in this example to get a feel for what happens if the sample size is not 'large enough.'

To summarize:

1. The sampling distributions of various statistics (the sampling distribution of the sample mean, or sample proportion, or sample count) are nearly normal. The normal distribution implies that a sample statistic that is close to the mean has a higher probability than one that is far away.
2. The mean of the sampling distribution of the sample mean is (in the limit) the same as the population mean.
3. It follows from the above two facts that the mean of a sample is more likely to be close to the population mean than not.

3.2 σ and $\sigma_{\bar{x}}$

We saw earlier that the standard deviation of the sampling distribution of the sample mean $\sigma_{\bar{x}}$ gets smaller as we increase sample size. When the sample has size 40, this standard deviation is 0.6156; when it is 100, this standard deviation is 0.4065.

Let's study the relationship between $\sigma_{\bar{x}}$ and σ. Recall that our population mean $\mu = 60$, $\sigma = 4$. The equation below summarizes the relationship; it shouldn't surprise you, since we just saw it above (also see Appendix A):

$$\sigma_{\bar{x}} = \frac{\sigma}{\sqrt{n}} \tag{3.2}$$

But note also that the tighter the distribution, the greater the probability that the estimate of the mean based on *a single sample* is close to the population mean. So the $\sigma_{\bar{x}}$ is an indicator of how good our estimate of the population mean is. As we increase the size of a single sample, the smaller the standard deviation of its corresponding sampling distribution becomes, and the higher the probability of its providing a good estimate of the population parameter. Let's quantify exactly how this estimate improves.

Histogram of sample.means

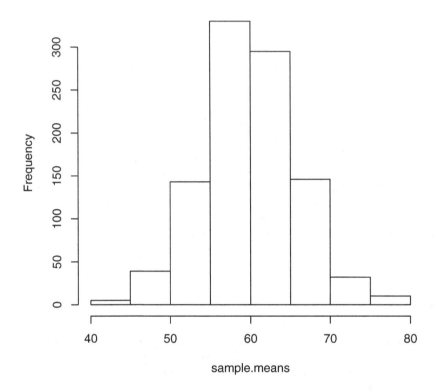

Fig. 3.5 The sampling distribution of sample mean from an exponentially distributed population.

3.3 The 95% Confidence Interval for the Sample Mean

Let's take a sample of 11 heights from a normally distributed population with known mean height $\mu = 60$ and SD $\sigma = 4$ (inches).

```
> sample.11 <- rnorm(11, mean = 60, sd = 4)

 [1] 58.88860 54.30641 59.35683 59.91140 56.21208 56.40233
 [7] 56.67023 63.84889 57.14325 50.89041 61.36143
```

And now let's estimate a population mean from this sample using the sample mean $\bar{x}$, and compute the SD $\sigma_{\bar{x}}$ of the corresponding sampling distribution. Since we know the precise population standard deviation we can get a precise value for $\sigma_{\bar{x}}$.

```
> estimated.mean <- mean(sample.11)
```

```
[1] 57.72653
```

```
> SD.population <- 4
```

```
[1] 4
```

```
> n <- length(sample.11)
```

```
[1] 11
```

```
> SD.distribution <- SD.population/sqrt(n)
```

```
[1] 1.206045
```

We know from the Central Limit Theorem that the sampling distribution of the sample mean is roughly normal, and we know that in this case $\sigma_{\bar{x}} = 1.2$. Recall that the probability that the population mean is within 2 $\sigma_{\bar{x}}$ of the sample mean is a bit over 0.95. Let's calculate this range:

$$\bar{x} \pm (2 \times \sigma_{\bar{x}}) = 58 \pm (2 \times 1.206) \tag{3.3}$$

The 0.95 probability region is between 55.3 and 60.1. The number 0.95 is a probability from the point of view of the sampling distribution, and a confidence level from the point of view of parameter estimation. In the latter case it's conventionally expressed as a percentage and is called the 95% confidence interval (CI).

Suppose now that sample size was four times bigger (44). Let's again calculate the sample mean, the standard deviation of the corresponding sampling distribution, and from this information, compute the 95% confidence interval.

```
> sample.44 <- rnorm(44, mean = 60, sd = 4)
> estimated.mean <- mean(sample.44)
> n <- length(sample.44)
> (SD.distribution <- 4/sqrt(n))
```

```
[1] 0.6030227
```

Now we get a much tighter 95% confidence interval:

$$\bar{x} \pm 2 \times \sigma_{\bar{x}} = 60 \pm 2 \times 0.603 \tag{3.4}$$

The interval now is between 59.2 and 61.6, smaller than the one we got for the smaller sample size. In fact, it is exactly half as wide. Take a moment to make sure you understand why.

3.4 Realistic Statistical Inference

Until now we have been sampling from a population whose mean and standard deviation we know. However, we normally don't know the population parameters. In other words, although we know that:

$$\sigma_{\bar{x}} = \frac{\sigma}{\sqrt{n}} \tag{3.5}$$

when we take samples in real life, we almost never know σ. After all, it is based on an average of distances from the population mean μ, and that is usually the very thing we are trying to estimate!

What we *do* have, however, is the standard deviation *of the sample itself* (denoted s). This in turn means that we can only get an *estimate* of $\sigma_{\bar{x}}$. This is called the STANDARD ERROR (SE) of the sample mean (or of whatever statistic we are measuring.):

$$SE_{\bar{x}} = \frac{s}{\sqrt{n}} \tag{3.6}$$

Pay careful attention to the distinction between s (an estimate of the standard deviation of the population σ) and $SE_{\bar{x}}$ (an estimate of the standard deviation of the sampling distribution, which is in turn based on s).

We saw previously that the size of $\sigma_{\bar{x}}$—a measure of the spread of the sampling distribution—is crucial in determining the size of a 95% confidence interval for a particular sample. Now we only have an estimate of that spread. Moreover, the estimate will change from sample to sample, as the value of s changes. This introduces a new level of uncertainty into our task: it has become an estimate based on an estimate! Intuitively, we would expect the confidence interval to increase in size, reflecting this increase in uncertainty. We will see how to quantify this intuition presently.

First, however, we should explore the pattern of variability in this new statistic we have introduced, s, which (like the sample mean) will vary randomly from sample to sample. Can we safely assume that s is a reliable estimate of σ?

3.5 s is an Unbiased Estimator of σ

Earlier in this chapter we repeatedly sampled from a population of people with mean age 60 and standard deviation 4; then we plotted the distribution of sample means that resulted from the repeated samples. One thing we noticed was that any one sample mean was more likely to be close to the population mean. Let's repeat this experiment, but this time we plot the distribution of the sample's standard deviation (Figure 3.6).

```
> sample.sds <- rep(NA, 1000)
> for (i in c(1:1000)) {
      sample.40 <- rnorm(40, mean = 60, sd = 4)
      sample.sds[i] <- sd(sample.40)
  }
> hist(sample.sds)
```

What we see is that any one sample's standard deviation *s* is more likely to be close to the population standard deviation σ. This is because the sampling distribution of the standard deviations also has a normal distribution.

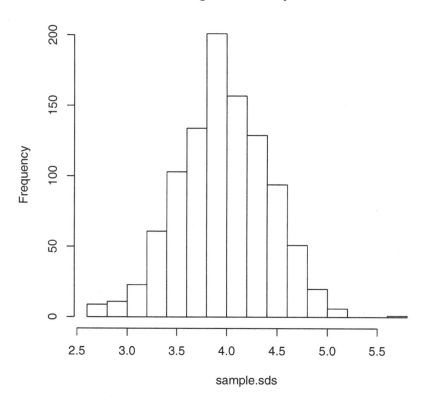

Fig. 3.6 The distribution of the sample standard deviation, sample size 40. The population is normally distributed.

So, if we use *s* as an estimator of σ we're more likely than not to get close to the right value: we say *s* is an UNBIASED ESTIMATOR of σ. This is true even if the population is not normally distributed. Let's check this again for an exponentially distributed population whose SD is 1 (Figure 3.7).

```
> sample.sds <- rep(NA, 1000)
> for (i in c(1:1000)) {
      sample.sds[i] <- sd(rexp(40))
  }
> hist(sample.sds)
```

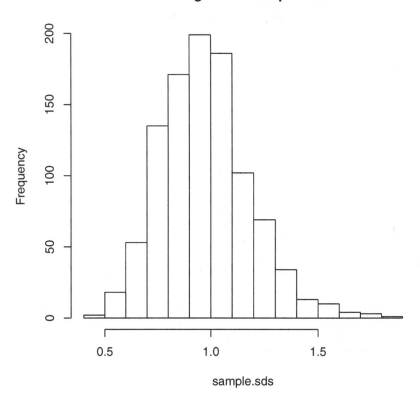

Fig. 3.7 The sampling distribution of sample standard deviations from an exponentially distributed population.

We are now at the point where we can safely use the sample standard deviation s as an estimate of the unknown population standard deviation σ, and this in turn allows us to estimate the standard deviation of the sampling distribution $\sigma_{\bar{x}}$ using the Standard Error $SE_{\bar{x}}$.

Notice that the Standard Error will vary from sample to sample, since the estimate *s* of the population parameter σ will vary from sample to sample. And of course, as the sample size increases the estimate *s* becomes more accurate, as does the SE, suggesting that the uncertainty introduced by this extra layer of estimation will be more of an issue for smaller sample sizes.

Our problem now is that the sampling distribution of the sample mean can no longer be modeled by the normal distribution, which is the distribution based ultimately on the *known* parameter σ.

If we were to derive some value *v* for the SE, and simply plug this in to the normal distribution for the sample statistic, this would be equivalent to claiming that *v* *really was* the population parameter σ.

What we require is a distribution whose shape has greater uncertainty built into it than the normal distribution. This is the motivation for using the so-called t-DISTRIBUTION, which we turn to next.

3.6 The t-distribution

As discussed above, the distribution needs to reflect greater uncertainty at small sample sizes. There is in fact a family of t-distribution curves whose shapes vary with sample size. In the limit, if the sample were the size of the entire population, the t-distribution would *be* the normal distribution (since then *s* would *be* σ), so the t-curve becomes more normal as sample size increases. This distribution is formally defined by the DEGREES OF FREEDOM (which is simply sample size minus 1 in this case) and has more of the total probability located in the tails of the distribution. It follows that the probability of a sample mean being close to the true mean is slightly lower when measured by this distribution, reflecting our greater uncertainty. You can see this effect in Figure 3.8 at small sample sizes:

```
> range <- seq(-4, 4, 0.01)
> multiplot(2, 2)
> for (i in c(2, 5, 15, 20)) {
        plot(range, dnorm(range), lty = 1, col = gray(0.5),
            xlab = "", ylab = "", cex.axis = 1.5)
        lines(range, dt(range, df = i), lty = 2,
            lwd = 2)
        mtext(paste("df=", i), cex = 1.2)
    }
```

But notice that with about 15 degrees of freedom, the t-distribution is already very close to normal.

What do we have available to us to work with now? We have an estimate *s* of the population SD, and so an estimate $SE_{\bar{x}}$ of the SD of the sampling distribution:

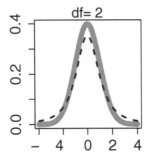
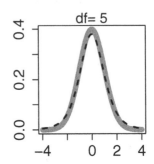

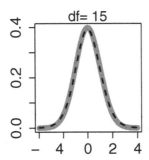
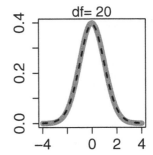

Fig. 3.8 A comparison between the normal (solid gray line) and t-distribution (broken black line) for different degrees of freedom.

$$SE_{\bar{x}} = \frac{s}{\sqrt{n}} \tag{3.7}$$

We also have a more spread-out distribution than the normal, the t-distribution, defined by the degrees of freedom (in this case, sample size minus 1). We are now ready to do realistic statistical inference.

3.7 The One-sample t-test

How do we build a confidence interval based on this new model of inference?

We need to ask: how many SE's do we need to go to the left and right of the sample mean, within the appropriate t-distribution, to be 95% sure that the population mean lies in that range? In the pre-computing days, people used to look up a table that told you, for $n-1$ degrees of freedom, how many SE's you need to go around the sample mean to get a 95% CI. Now we can ask R. First we take a sample of size 11 from a population with mean 60 and standard deviation 4.

```
> sample <- rnorm(11, mean = 60, sd = 4)
```

Using this sample, we can ask for the 95% confidence interval:

```
> t.test(sample)$conf.int
```

```
[1] 56.78311 63.56462
attr(,"conf.level")
[1] 0.95
```

Note that all of the information required to perform this inference is contained in the sample itself: the sample mean; the sample size and sample standard deviation s (from which we compute the SE), the degrees of freedom (the sample size minus 1, from which we reference the appropriate t-distribution). Sure enough, if our sample size had been larger, our CI would be narrower:

```
> sample <- rnorm(100, mean = 60, sd = 4)
```

```
> t.test(sample)$conf.int
```

```
[1] 58.62889 60.20964
attr(,"conf.level")
[1] 0.95
```

3.8 Some Observations on Confidence Intervals

There are some subtleties associated with confidence intervals that are often not brought up in elementary discussions, simply because the issues are just too daunting to tackle. However, we will use simulations to unpack some of these subtleties. We hope that the reader will see that the issues are in reality quite simple.

The first critical point to understand is the meaning of the confidence interval. We have been saying up till now that the 95% confidence interval tells you the range within which we are 95% sure that the population mean lies. However, one important point to notice is that the range defined by the confidence interval will vary with each sample even if the sample size is kept constant. The reason is that the sample mean will vary each time, and the standard deviation will vary too. We can check this fact quite easily.

First we define a function for computing 95% CIs:[1]

```
> se <- function(x) {
      y <- x[!is.na(x)]
      sqrt(var(as.vector(y))/length(y))
  }
> ci <- function(scores) {
      m <- mean(scores, na.rm = TRUE)
      stderr <- se(scores)
      len <- length(scores)
      upper <- m + qt(0.975, df = len - 1) * stderr
      lower <- m + qt(0.025, df = len - 1) * stderr
      return(data.frame(lower = lower, upper = upper))
  }
```

Next, we take 100 samples repeatedly from a population with mean 60 and SD 4, computing the 95% CI each time.

```
> lower <- rep(NA, 100)
> upper <- rep(NA, 100)
> for (i in 1:100) {
      sample <- rnorm(100, mean = 60, sd = 4)
      lower[i] <- ci(sample)$lower
      upper[i] <- ci(sample)$upper
  }
> cis <- cbind(lower, upper)
> head(cis)

          lower     upper
[1,]  59.30995  60.84080
[2,]  59.59644  61.10296
[3,]  58.90593  60.61683
[4,]  58.51755  60.22174
[5,]  58.85010  60.50782
[6,]  59.44144  60.87652
```

Thus, the center and the size of any one confidence interval, based on a single sample, will depend on the particular sample mean and standard deviation you happen to observe for that sample. The sample mean and standard deviation are likely to be close to the population mean and standard deviation, but they are ultimately just estimates of the true parameters.

Importantly, because of the normally distributed shapes of the distribution of sample means and sample standard deviations (see Figures 3.5 and 3.7), if we repeatedly sample from a population and compute the confidence intervals

---

[1] Here, we use a built-in R function called qt(p,DF) which, for a given confidence-interval range (say, 0.975), and a given degrees of freedom, DF, tells you the corresponding critical t-value.

each time, in approximately 95% of the confidence intervals the population
mean will lie within the ranges specified. In the other 5% or so of the cases,
the confidence intervals will not contain the population mean.

This is what the '95%' confidence interval means: it's a statement about
hypothetical repeated samples. More specifically, it's a statement about the
probability that the hypothetical confidence intervals (that would be computed from the hypothetical repeated samples) will contain the population
mean.

Let's check the above statement. We can repeatedly build 95% CIs and
determine whether the population mean lies within them. The claim is that
it will lie within the CI approximately 95% of the time.

```
> store <- rep(NA, 100)
> pop.mean <- 60
> pop.sd <- 4
> for (i in 1:100) {
        sample <- rnorm(100, mean = pop.mean, sd = pop.sd)
        lower[i] <- ci(sample)$lower
        upper[i] <- ci(sample)$upper
        if (lower[i] < pop.mean & upper[i] > pop.mean) {
            store[i] <- TRUE
        }
        else {
            store[i] <- FALSE
        }
    }
> cis <- cbind(lower, upper)
> store <- factor(store)
> summary(store)

FALSE   TRUE
    6     94
```

So that's more or less true. To drive home the point, we can also plot
the confidence intervals to visualize the proportion of cases where each CI
contains the population mean (Figure 3.9).

```
> main.title <- "95% CIs in 100 repeated samples"
> line.width <- ifelse(store == FALSE, 2, 1)
> cis <- cbind(cis, line.width)
> x <- 0:100
> y <- seq(55, 65, by = 1/10)
> plot(x, y, type = "n", xlab = "i-th repeated sample",
        ylab = "Scores", main = main.title)
> abline(60, 0, lwd = 2)
> x0 <- x
> x1 <- x
```

```
> arrows(x0, y0 = cis[, 1], x1, y1 = cis[, 2],
        length = 0, lwd = cis[, 3])
```

In this figure, we control the width of the lines marking the CI using the information we extracted above (in the object `store`) to determine whether the population mean is contained in the CI or not: when a CI does not contain the population mean, the line is thicker than when it does contain the mean. You should try repeatedly sampling from the population as we did above, computing the lower and upper ranges of the 95% confidence interval, and then plotting the results as shown in Figure 3.9.

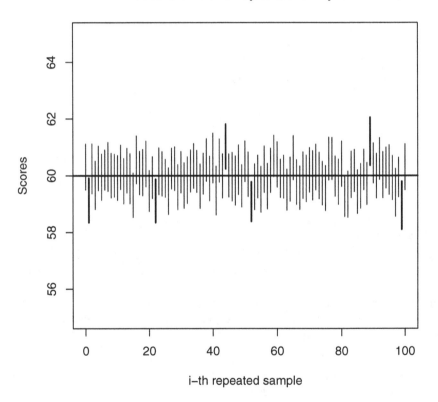

Fig. 3.9 A visualization of the proportion of cases where the population mean is contained in the 95% CI, computed from repeated samples. The CIs that do not contain the population mean are marked with thicker lines.

Note that when we compute a 95% confidence interval for a particular sample, we have only *one* interval. Strictly speaking, that particular interval does *not* mean that the probability that the population mean lies *within that interval* is 0.95. For that statement to be true, it would have to be the case that the population mean is a random variable, like the heads and tails in a coin are random variables, and 1 through 6 on a die are random variables.

The population mean is a single point value that cannot have a multitude of possible values and is therefore not a random variable. If we relax this assumption, that the population mean is a point value, and assume instead that 'the' population mean is in reality a range of possible values (each value having different probabilities of being the population mean), then we could say that any one 95% confidence interval represents the range within which the population mean lies with probability 0.95. See the book by Gelman and Hill (2007) for more detail on this approach.

It's worth repeating the above point about confidence intervals. The meaning of the confidence interval depends crucially on hypothetical repeated samples: the confidence intervals computed in 95% of these repeated samples will contain the population mean. In essence, the confidence interval from a single sample is a random variable just like heads and tails in a coin toss, or the numbers 1 through 6 in a die, are random variables. Just as a fair coin has a 0.5 chance of yielding a heads, and just as a fair die has a 1/6 chance of landing a 1 or 2 etc., a confidence interval in repeated sampling has a 0.95 chance of containing the population mean.

3.9 Sample SD, Degrees of Freedom, Unbiased Estimators

Let's revisit the question: Why do we use $n-1$ in the equation for standard deviation? Recall that the sample standard deviation s is just the root of the variance: the average distance of the numbers in the list from the mean of the numbers:

$$s^2 = \frac{1}{n-1}\sum_{i=1}^{n}(x_i - \bar{x})^2 \qquad (3.8)$$

We can explore the reason why we use $n-1$ in the context of estimation by considering what would happen if we simply used n instead. As we will see, if we use n, then s (which is an estimate of the population variance σ) would be smaller. This smaller s turns out to provide a poorer estimate than when we use $n-1$: it is a BIASED ESTIMATOR. Let's verify this using simulations.

We define new variance and standard deviation functions that use n, and simulate the sampling distribution of this new statistic s' from a population with known standard deviation $\sigma = 1$).

```
> new.var <- function(x) {
      sum((x - mean(x))^2)/length(x)
  }
```

```
> new.sd <- function(x) {
        sqrt(new.var(x))
   }
> correct <- rep(NA, 1000)
> incorrect <- rep(NA, 1000)
> for (i in 1:1000) {
        sample.10 <- rnorm(10, mean = 0, sd = 1)
        correct[i] <- sd(sample.10)
        incorrect[i] <- new.sd(sample.10)
   }
```

As shown below (Figure 3.10), using n gives a biased estimate of the true standard deviation:

```
> multiplot(1, 2)
> hist(correct, main = paste("Mean ", round(mean(correct),
        digits = 2), sep = " "))
> hist(incorrect, main = paste("Mean ", round(mean(incorrect),
        digits = 2), sep = " "))
```

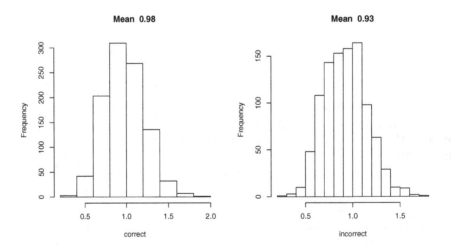

Fig. 3.10 The consequence of taking $n-1$ versus n in the denominator for calculating variance, sample size 10.

3.10 Summary of the Sampling Process

It is useful at this point to summarize the terminology we have been using, and the logic of sampling. First, take a look at the concepts we have covered so far.

We provide a list of the different concepts in Table 3.1 below for easy reference. Here, $\sigma_{\bar{x}} = \frac{\sigma}{\sqrt{n}}$ and $SE_{\bar{x}} = \frac{s}{\sqrt{n}}$.

Table 3.1 A summary of the notation used.

The sample statistic	is an unbiased estimate of
sample mean $\bar{x}$	population mean μ
sample SD s	population SD σ
standard error $SE_{\bar{x}}$	sampling distribution $\sigma_{\bar{x}}$

1. Statistical inference involves a single sample value but assumes knowledge of the sampling distribution which provides probabilities for all possible sample values.
2. The statistic (e.g., mean) in a random sample is more likely to be closer to the population parameter (the population mean) than not. This follows from the normal distribution of the sample means.
3. In the limit, the mean of the sampling distribution is equal to the population parameter.
4. The further away a sample statistic is from the mean of the sampling distribution, the lower the probability that such a sample will occur.
5. The standard deviation of the sampling distribution $\sigma_{\bar{x}}$ is partially determined by the inherent variability σ in the population, and partially determined by the sample size. It tells us how steeply the probability falls off from the center. If $\sigma_{\bar{x}}$ is small, then the fall-off in probability is steep: random samples are more likely to be very close to the mean, samples are better indicators of the population parameters, and inference is more certain. If $\sigma_{\bar{x}}$ is large, then the fall-off in probability from the center is gradual: random samples far from the true mean are more likely, samples are not such good indicators of the population parameters, and inference is less certain.
6. While we do not know $\sigma_{\bar{x}}$, we can estimate it using $SE_{\bar{x}}$ and perform inference using a distribution that is almost normal, but reflects the increase in uncertainty arising from this estimation: the t-distribution.

3.11 Significance Tests

Recall the discussion of 95% confidence intervals: The sample gives us a mean $\bar{x}$. We compute $SE_{\bar{x}}$ (an estimate of $\sigma_{\bar{x}}$) using s (an estimate of σ) and sample size n. Then we calculate the range $\bar{x} \pm 2 \times SE_{\bar{x}}$. That's the 95% CI.

We don't know the population mean—if we did, why bother sampling? But suppose we have a *hypothesis* that the population mean has a certain value. If we have a hypothesis about the population mean, then we also know what the corresponding sampling distribution would look like: we know the probability of any possible sample given that hypothesis. We then take an actual sample, measure the distance of our sample mean from the hypothesized population mean, and use the facts of the sampling distribution to determine the probability of obtaining such a sample *assuming the hypothesis is true*. This amounts to a *test* of the hypothesis. Intuitively, if the probability of our sample (given the hypothesis) is high, this provides evidence the hypothesis is true. In a sense, this is what our hypothesis predicts. Conversely, if the probability of the sample is low (given the hypothesis), this is evidence against the hypothesis. The hypothesis being tested in this way is termed the NULL HYPOTHESIS. Let's do some simulations to better understand this concept.

Suppose our hypothesis, based perhaps on previous research, is that the population mean is 70, and let's assume for the moment the population $\sigma = 4$. This in turn means that the sampling distribution of the mean, given some sample size, say 11, would have a mean of 70, and a standard deviation $\sigma_{\bar{x}} = 1.2$:

```
> SD.distribution = 4/sqrt(11)

[1] 1.206045
```

Figure 3.11 shows what we expect our sampling distribution to look like if our hypothesis *were in fact* true. This hypothesized distribution is going to be our reference distribution on which we base our test.

```
> range <- seq(55, 85, 0.01)
> plot(range, dnorm(range, mean = 70, sd = SD.distribution),
       type = "l", ylab = "", main = "The null hypothesis")
```

Suppose now that we take an actual sample of 11 from a population whose mean μ is in fact (contra the hypothesis) 60:

```
> sample <- rnorm(11, mean = 60, sd = 4)

> sample.mean <- mean(sample)

[1] 60.76659
```

Inspection of (Figure 3.11) shows that, in a world in which the population mean was really 70, the probability of obtaining a sample whose mean is 61 is extremely low. Intuitively, this sample is evidence against the null hypothesis.

The null hypothesis

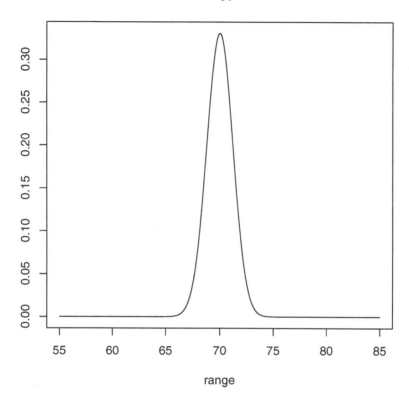

Fig. 3.11 A sampling distribution with mean 70 and $\sigma_{\bar{x}} = 1.206$.

A SIGNIFICANCE TEST provides a formal way of quantifying this insight. The result of such a test yields a probability that indicates exactly how well or poorly the data and the hypothesis agree.

3.12 The Null Hypothesis

While this perfectly symmetrical, intuitive way of viewing things ('evidence for', 'evidence against') is on the right track, there is a further fact about the null hypothesis which gives rise to an asymmetry in the way we perform significance tests.

The statement being tested in a significance test— the NULL HYPOTHE-SIS, H_0— is usually formulated in such a way that the statement represents 'no effect,' 'pure chance' or 'no significant difference'—or as Polya called it 'chance, the ever present rival conjecture' (Polya, 1954, 55). Scientists are generally not interested in proving 'no effect.' This is where the asymmetry comes in: we are usually not interested in finding evidence *for* the null hypothesis, conceived in this way. Rather, we are interested in evidence *against* the null hypothesis, since this is evidence for some real statistically significant result. This is what a formal significance test does: it determines if the result provides sufficient evidence against the null hypothesis for us to reject it. Note that if it doesn't provide sufficient evidence, we have *not* proved the contrary—we have not 'proved the null hypothesis.' We simply don't have enough evidence, *based on this single result*, to reject it. We sharpen this idea in chapter 4.

In order to achieve a high degree of skepticism about the interpretation of the data, we require the evidence against the null hypothesis to be very great. In our current example, you might think the result we obtained, 61, was fairly compelling evidence against it. But how do we quantify this? Intuitively, the further away from the mean of the sampling distribution our data lies, the greater the evidence against it. Statistically significant results reside out in the tails of the distribution. How far out? The actual values and ranges of values we get will vary from experiment to experiment, and statistic to statistic. How can we determine a general rule?

3.13 z-scores

We have already seen that, in a normal distribution, about 95% of the total probability falls within 2 SD of the mean, and thus 5% of the probability falls far out in the tails. One way of setting a general rule then, is to say that if an observed value falls far out in the tail of the distribution, we will consider the result extreme enough to reject the null hypothesis (we can set this threshold anywhere we choose: 95% is a conventional setting).

Recall our model: we know the sampling distribution we would see in a world in which the null hypothesis is true, in which the population mean is really 70 (and whose population σ is known to be 4). We also know this distribution is normal. How many SDs from the mean is our observation? Is it more than 2 SDs?

We need to express the difference between our observation $\bar{x}$ and hypothesized mean of the distribution μ_0 in units of the standard deviation of the distribution: i.e., some number z times $\sigma_{\bar{x}}$. We want to know this number z.

$$\bar{x} - \mu_0 = z\sigma_{\bar{x}} \tag{3.9}$$

Solving for z:

$$z = \frac{\bar{x} - \mu_0}{\sigma_{\bar{x}}} \tag{3.10}$$

$$= \frac{\bar{x} - \mu_0}{\sigma/\sqrt{n}} \tag{3.11}$$

z is called the STANDARDIZED VALUE or the Z-SCORE. In addition, one could imagine computing this standardized version of the sample mean every time we take a sample, in which case we have effectively defined a new statistic. Viewed in this way, the score is also referred to as a TEST-STATISTIC.

Let's make this concrete. Suppose in our current simulation we draw a sample whose mean is precisely 60: then $\bar{x} = 60, \mu_0 = 70, \sigma = 4, n = 11$. So we get:

$$z = \frac{60 - 70}{4/\sqrt{11}} \tag{3.12}$$

$$= -8.291562 \tag{3.13}$$

We see that this observation is well beyond 2 SDs from the mean, and thus represents statistically significant evidence against the null hypothesis.

Notice there is a natural interpretation of this process in terms of confidence intervals. We know that, on repeated sampling from a population that has mean 70 (the null hypothesis), in 95% of the samples the true population mean (70) would fall within two SDs of the sample mean (which of course would be different for each sample). It follows that in 5% of the samples, this 2 SD confidence interval will *not* contain the population mean. These are precisely the samples which fall out in the tails of the distribution, more than 2 SD from the distribution mean.

z-scores are a quick and accepted way of expressing 'how far away' from the hypothetical value an observation falls, and for determining if that observation is beyond some accepted threshold. Ultimately, however, they take their meaning from the probability corresponding to the value, which is traditionally expressed by rules-of-thumb (2 SD corresponds to 95%), or tables which translate particular scores to particular probabilities. It is this probability we turn to next.

3.14 P-values

It would be nice if we could set a probability threshold like this: 'If the probability of the result is less than 0.05, given the null hypothesis, then we reject the null hypothesis.' But we can't. Recall from Chapter 2 that the probability of obtaining *any particular result* out of all of the possibilities is very low (see page 25), and varies with the number of possibilities under consideration. This value does not generalize from experiment to experiment.

Although we cannot use the actual probability of the observed value, we can usefully ask *how much of the total probability lies beyond the observed value*, out into the tail of the distribution. In the discrete case (binomial distribution) this is a sum of probabilities, in the continuous case (normal distribution) an area under the curve. Call o_1 the observed value, o_2 the next value out, o_3 the next, and so on until we exhaust all the possibilities. The sum of these is the probability of a complex event, the probability of 'observing the value o_1 or a value more extreme.' (Once again, we couch our probability measure in terms of a range of values). This then is a measure, based directly on probability, of 'how far away' from the mean an observed value lies. The smaller this probability, the more extreme the value. We can now say, if this probability is less than 0.05, we reject the hypothesis. The technical name for this measure is the P-VALUE.

In short, the p-value of a statistical test is the probability, computed assuming that H_0 is true, that the test statistic would take a value as extreme or more extreme than that actually observed.

Note that this is a CONDITIONAL PROBABILITY: it is the probability of observing a particular sample mean (or something more extreme) conditional on the assumption that the null hypothesis is true. We can write this conditional probability as $P(\text{Observed mean} \mid H_0)$, or even more succinctly as $P(\text{Data} \mid H_0)$. The p-value does *not* measure the probability of the null hypothesis given the data, $P(H_0 \mid \text{Data})$. There is a widespread misunderstanding that the p-value tells you the probability that the null hypothesis is true (in light of some observation); it doesn't. You can confirm easily that we cannot switch conditional probabilities. The probability of the streets being wet given that rain has fallen $P(\text{Wet Streets} \mid \text{Rain})$ (presumably close to 1) is not at all the same as the probability of rain having fallen given that the streets are wet $P(\text{Rain} \mid \text{Wet Streets})$. There are many reasons why the streets may be wet (street cleaning, burst water pipes, etc.), rain is just one of the possibilities. There is however a technical term that expresses the intuition about the hypothesis in light of the data: if $P(\textit{Data} \mid H_0)$ is low, we say the LIKELIHOOD of the hypothesis is low.

How do we determine this p-value? We simply integrate the area under the normal curve, going out from our observed value. (Recall that, for the present, we are assuming we *know* the population parameter σ, so it is appropriate to use the normal distribution). We actually have two completely equivalent ways to do this, since we now have two values (the actual observed value and its z-score), and two corresponding curves (the sampling distribution where the statistic is the sample mean, and the sampling distribution where the statistic is the standardized mean, the 'z-statistic'). We have seen what the sampling distribution of the sample mean looks like, assuming the null hypothesis is true (i.e. $\mu_0 = 70$, Figure 3.11). What is the sampling distribution of the z-statistic under this hypothesis? Let's do a simulation to find out.

In Figure 3.12, we repeat the simulation of sample means that we carried out at the beginning of the chapter, but now using the parameters of our current null hypothesis $\mu_0 = 70$, $\sigma = 4$, sample size $= 11$. But in addition, for each sample we also compute the z-statistic, according to the formula provided above. We also include the corresponding normal curves for reference (recall these represent the limiting case of the simulations). As you can see, the distribution of the z-statistic is normal, with mean $= 0$, and SD $= 1$. A normal distribution with precisely these parameters is known as the STANDARDIZED NORMAL DISTRIBUTION.

```
> sample.means <- rep(NA, 1000)
> zs <- rep(NA, 1000)
> for (i in 1:1000) {
        sample.11 <- rnorm(11, mean = 70, sd = 4)
        sample.means[i] <- mean(sample.11)
        zs[i] <- (mean(sample.11) - 70)/(4/sqrt(11))
   }
> multiplot(2, 2)
> sd.dist <- 4/sqrt(11)
> plot(density(sample.means), xlim = range(70 -
        (4 * sd.dist), 70 + (4 * sd.dist)), xlab = "",
        ylab = "", main = "")
> plot(density(zs), xlim = range(-4, 4), xlab = "",
        ylab = "", main = "")
> plot(function(x) dnorm(x, 70, 4/sqrt(11)), 70 -
        (4 * sd.dist), 70 + (4 * sd.dist), xlab = "",
        ylab = "", main = "")
> plot(function(x) dnorm(x, 0, 1), -4, 4, xlab = "",
        ylab = "", main = "")
```

The crucial thing to note is that the area from either value out to the edge, which is the probability of interest, is precisely the same in the two cases, so we can use either. It is traditional to work with the standardized values, for reasons that will become clear.

Recall the z-score for our actual observation was -8.291562. This is an extreme value, well beyond 2 SDs from the mean, so we would expect there to be very little probability between it and the left tail of the distribution. We can calculate it directly by integration:

```
> integrate(function(x) dnorm(x, mean = 0, sd = 1),
        -Inf, -8.291562)

5.588542e-17 with absolute error < 4.5e-24
```

This yields a vanishingly small probability. We also get precisely the same result using the actual observed sample mean with the original sampling distribution:

```
> integrate(function(x) dnorm(x, mean = 70, sd = 4/sqrt(11)),
        -Inf, 60)
```

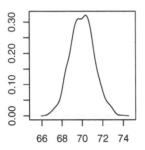 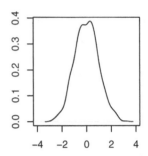

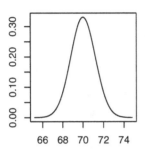 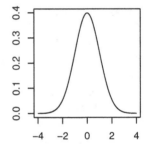

Fig. 3.12 The sampling distribution of the sample mean (left) and its z-statistic (right).

```
5.588543e-17 with absolute error < 6.2e-20
```

Suppose now we had observed a sample mean of 67.58. This is much closer to the hypothetical mean of 70. The standardized value here is almost exactly -2.0:

```
> (67.58 - 70)/(4/sqrt(11))
```

```
[1] -2.006558
```

Integrating under the standardized normal curve we find the following probability:

```
> integrate(function(x) dnorm(x, 0, 1), -Inf, -2)
```

```
0.02275013 with absolute error < 1.5e-05
```

This accords well with our rule-of-thumb. About 95% of the probability is within 2 SD of the mean. The remainder is split into two, one at each end of the distribution, each representing a probability of about 0.025 (actually a little less).

3.15 Hypothesis Testing: A More Realistic Scenario

In the above example we were able to use the standard deviation of the sampling distribution $\sigma_{\bar{x}}$, because we were given the standard deviation of the population σ. As we remarked earlier, in the real world we usually do not know σ, it's just another unknown parameter of the population. Just as in the case of computing real world confidence intervals, instead of σ we use the unbiased estimator s; instead of $\sigma_{\bar{x}}$ we use the unbiased estimator $SE_{\bar{x}}$; instead of the normal distribution we use the t-distribution.

Recall the z-score:

$$z = \frac{\bar{x} - \mu_0}{\sigma_{\bar{x}}} \tag{3.14}$$

$$= \frac{\bar{x} - \mu_0}{\sigma/\sqrt{n}} \tag{3.15}$$

And recall our formal definition of a statistic: a number that describes some aspect of the sample. Using this definition, the z-score seems to fail as a statistic, since it makes reference to a population *parameter* σ. But if we now replace that parameter with an estimate s derived from the sample itself, we get the so-called t-statistic:

$$t = \frac{\bar{x} - \mu_0}{SE_{\bar{x}}} \tag{3.16}$$

$$= \frac{\bar{x} - \mu_0}{s/\sqrt{n}} \tag{3.17}$$

This then can also be interpreted as yet another sampling statistic, with its own distribution. What does the distribution for *this* statistic look like? This is exactly the question William Gossett posed in 1908, in the paper that introduced the t-distribution (Gossett published under the pseudonym 'Student', and the statistic is often referred to as Student's t). Interestingly, before he was able to develop a precise mathematical formulation, he used simulation to get a feel for the distribution, as he explains (Student, 1908, 13):

> Before I had succeeded in solving my problem analytically, I had endeavoured to do so empirically. The material used was a correlation table containing the height and left middle finger measurements of 3000 criminals [...] The mea-

surements were written out on 3000 pieces of cardboard, which were then very thoroughly shuffled and drawn at random. As each card was drawn its numbers were written down in a book which thus contains the measurements of 3000 criminals in a random order. Finally each consecutive set of 4 was taken as a sample...

In the spirit of Student then, let's simulate the sampling distribution of the t-statistic and compare it to the distribution of the z-statistic shown above (Figure 3.13). Like Student, we use a small sample size of 4, where the effect of estimating σ by s is more noticeable. And again, we include the limiting-case normal and t-curves for reference. Notice the subtle but real effect of an increase in probability in the tails of the t-distribution. More of the probability is located further from the mean, reflecting the increase in uncertainty due to the fact that we are *estimating* the standard deviation of the population.

```
> zs <- rep(NA, 10000)
> ts <- rep(NA, 10000)
> for (i in 1:10000) {
        sample.4 <- rnorm(4, mean = 70, sd = 4)
        zs[i] <- (mean(sample.4) - 70)/(4/sqrt(4))
        ts[i] <- (mean(sample.4) - 70)/(sd(sample.4)/sqrt(4))
  }
> multiplot(2, 2)
> plot(density(zs), xlim = range(-4, 4), xlab = "z-scores",
        ylab = "", main = "Sampling distribution of z")
> plot(density(ts), xlim = range(-4, 4), xlab = "t-scores",
        ylab = "", main = "Sampling distribution of t")
> plot(function(x) dnorm(x, 0, 1), -4, 4, xlab = "x",
        ylab = "", main = "Limiting case: normal distribution")
> plot(function(x) dt(x, 3), -4, 4, xlab = "x",
        ylab = "", main = "Limiting case: t-distribution")
```

As discussed earlier, there is a family of t-curves corresponding to the different sample sizes; the t-curve is almost identical to the standardized normal curve, especially at larger sample sizes.

Note a rather subtle point: we can have samples with the same mean value, 67.58 say, but different t-scores, since the SD s of the samples may differ. In order to facilitate comparison with the z-score situation, let's suppose that we just happen to find a sample whose SD s is identical to the population SD $\sigma = 4$. In just such a case the t-score would be identical to the z-score, but the probability associated with the score will differ slightly, since we use the t-distribution, not the normal distribution.

The t-score here will be:

```
> (67.58 - 70)/(4/sqrt(11))

[1] -2.006558
```

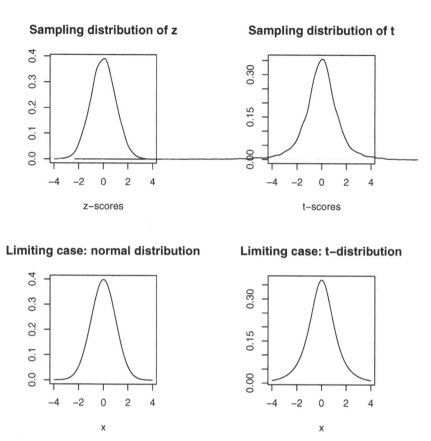

Fig. 3.13 The sampling distribution of the z-statistic (left) and the t-statistic (right).

And so for a sample size of 11 (degrees of freedom 10) we have the corresponding probability:

```
> integrate(function(x) dt(x, 10), -Inf, -2.006558)
```

```
0.03629508 with absolute error < 3.8e-06
```

Notice this probability is 'less extreme' than when we use the normal curve. The evidence against the null hypothesis is not as strong, reflecting the fact that we are uncertain of the true SD of the sampling distribution. This is the uncertainty built into the t-curve, with more of its probability in the tails.

Once again we note that, as sample size increases, the estimate s improves, and the more our results resemble the normal curve:

```
> integrate(function(x) dt(x, 20), -Inf, -2.006558)
```

```
0.02925401 with absolute error < 3.8e-06
```

Note that our null hypothesis H_0 was that the observed mean $\bar{x}$ is equal to the hypothesized mean μ_0. Thus, rejecting the null hypothesis amounts to accepting the alternative hypothesis, i.e., that the observed value is less than the mean *or* the observed value is greater than the mean:

$$H_a : \bar{x} < \mu_0 \text{ or } \mu_0 < \bar{x} \tag{3.18}$$

This means that as evidence for rejection of H_0 we will count extreme values on *both* sides of μ. For this reason, the above test is called a TWO-SIDED SIGNIFICANCE TEST (also known as the TWO-TAILED SIGNIFICANCE TEST). Note that if we simply reported the probability corresponding to the t-value t, we would *not* be reporting the probability of 'a value being more than t away' from the mean, but the probability in one direction only. For that reason, in a two-sided test, since the distributions are symmetrical, the p-value will be twice the value of the probability corresponding to the particular t-value we obtain. If the p-value is $\leq \alpha$, we say that the data are significant at level α. Purely by convention, $\alpha = 0.05$.

By contrast, if our null hypothesis were that the observed mean $\bar{x}$ is, say, equal to or less than the hypothesized mean μ_0, then the alternative hypothesis would be:

$$H_a : \mu_0 < \bar{x} \tag{3.19}$$

In this situation, we would use a one-sided significance test, reporting the probability in the relevant direction only.

R does everything required for a t-test of significance as follows, and you can specify (inter alia) what your μ_0 is (note that it need not be zero), whether it is two-sided or not (see the documentation for the `t.test` for details on how to specify this), and the confidence level (the α level) you desire, as follows:

```
> sample.11 <- rnorm(11, mean = 60, sd = 4)
> t.test(sample.11, alternative = "two.sided",
        mu = 70, conf.level = 0.95)

        One Sample t-test

data:  sample.11
t = -8.8974, df = 10, p-value = 4.587e-06
alternative hypothesis: true mean is not equal to 70
95 percent confidence interval:
 56.67570 62.01268
sample estimates:
mean of x
  59.3442
```

Experiment with the above code: change the hypothetical mean, change the mean of the sampled population and its SD, change the sample size, etc. In each case, see how the sample mean, the t-score, the p-value and the confidence interval differ.

It is also instructive to keep the parameters the same and simply repeat the experiment, taking different random samples each time (effectively, REPLICATING the experiment). Watch how the p-values change, watch how they change from replicate to replicate under different parameter settings. Do you ever find you would accept the null hypothesis when it is in fact false? How likely is it that you would make a mistake like that? This is an issue we will return to in more depth in Chapter 4.

The t-value we see above is indeed the t in equation 3.16; we can verify this by doing the calculation by hand:

```
> (mean(sample.11) - 70)/se(sample.11)

[1] -8.897396
```

3.16 Comparing Two Samples

In one-sample situations our null hypothesis is that there is no difference between the sample mean and the population mean:

$$H_0 : \bar{x} = \mu \tag{3.20}$$

When we compare samples from two different populations, we ask the question: are the population means identical or not? Our goal now is to figure out some way to define our null hypothesis in this situation.

Consider this example of a common scenario in experimental research. Mean reading times and standard deviations are available for children and adults reading English sentences. Let us say that we want to know whether children are faster or slower than adults in terms of reading time. You probably don't need to do an experiment to answer this question, but it will do as an illustration of this type of experiment.

We know that, due to the nature of random sampling, there is bound to be *some* difference in sample means even if the population means are identical. We can reframe the research question as follows: is the difference observed in the two sample means a true difference or just a chance event? The data are shown in Table 3.2.

Notice a few facts about the data. We have different sample sizes in each case. How will that affect our analysis? Notice too that we have different standard deviations in each case: this makes sense, since children exhibit a wider range of abilities than literate adults. But we now know how great an effect the variability of the data has on statistical inference. How will

Table 3.2 Hypothetical data showing reading times for adults and children.

group	sample size n	$\bar{x}$ (secs)	s
children	$n_1 = 10$	$\bar{x}_1 = 30$	$s_1 = 43$
adults	$n_2 = 20$	$\bar{x}_2 = 7$	$s_2 = 25$

we cope with these different SD's? Finally, the mean reading times certainly 'look' significantly different, but are we sure the difference is beyond the realm of chance, and if so, can we say exactly how much?

Such research problems have the properties that (i) the goal is to compare the responses in two groups; (ii) each group is considered a sample from a distinct population (a 'between-subjects' design); (iii) the responses in each group are independent of those in the other group; and (iv) the sample sizes of each group can be different.

The question now is, how can we formulate the null hypothesis?

3.16.1 H_0 in Two-sample Problems

Let us start by saying that the unknown population mean of children is μ_1, and that of adults is μ_2. We can state our null hypothesis as follows:

$$H_0 : \mu_1 = \mu_2 \qquad (3.21)$$

Equivalently, we can say that our null hypothesis is that the difference between the two means is zero:

$$H_0 : \mu_1 - \mu_2 = 0 = \delta \qquad (3.22)$$

We have effectively created a new population parameter δ:

$$H_0 : \delta = 0 \qquad (3.23)$$

We can now define a new statistic $d = \bar{x}_1 - \bar{x}_2$ and use that as an estimate of δ, which we've hypothesized to be equal to zero. But to do this we need a sampling distribution of the difference of the two sample means $\bar{x}_1$ and $\bar{x}_2$.

Let's do a simulation to get an understanding of this approach. For simplicity we will use the sample means and standard deviations from the example above as our population parameters in the simulation, and we will also use the sample sizes above for the repeated sampling. Assume a population with $\mu_1 = 30$, $\sigma_1 = 43$, and another with mean $\mu_2 = 7$, $\sigma_2 = 25$. So we already know in this case that the null hypothesis is false, since $\mu_1 \neq \mu_2$. But let's take 1000 sets of samples of each population, compute the differences in mean in each set of samples, and plot that distribution of *the differences of the sample mean*:

```
> d <- rep(NA, 1000)
> for (i in 1:1000) {
        sample1 <- rnorm(10, mean = 30, sd = 43)
        sample2 <- rnorm(20, mean = 7, sd = 25)
        d[i] <- mean(sample1) - mean(sample2)
  }
```

Note that the mean of the differences-vector **d** is close to the true difference:

```
> 30 - 7
```

[1] 23

```
> mean(d)
```

[1] 23.09159

Then we plot the distribution of **d**; we see a normal distribution (Figure 3.14).

```
> hist(d)
```

So, the distribution of the differences between the two sample means is normally distributed, and centered around the true difference between the two populations. It is because of these properties that we can safely take *d* to be an unbiased estimator of δ. How accurate an estimator is it? In other words, what is the standard deviation of this new sampling distribution? It is clearly dependent on the standard deviation of the two populations in some way:

$$\sigma_{\bar{x}_1 - \bar{x}_2} = f(\sigma_1, \sigma_2) \tag{3.24}$$

(Try increasing one or other or both of the σ in the above simulation to see what happens). The precise relationship is fundamentally additive: instead of taking the root of the variance, we take the root of the sum of variances:

$$\sigma_{\bar{x}_1 - \bar{x}_2} = \sqrt{\frac{\sigma_1^2}{n_1} + \frac{\sigma_2^2}{n_2}} = \sqrt{\frac{43^2}{10} + \frac{25^2}{20}} = 14.702. \tag{3.25}$$

```
> newsigma <- round(sqrt((43^2/10) + (25^2/20)),
        digits = 4)
```

In our single sample, $\bar{x}_1 - \bar{x}_2 = 17$. The null hypothesis is $\mu_1 - \mu_2 = 0$. How should we proceed? Is this sample difference sufficiently far away from the hypothetical difference (0) to allow us to reject the null hypothesis? Let's first translate the observed difference 17 into a z-score. Recall how the z-score is calculated:

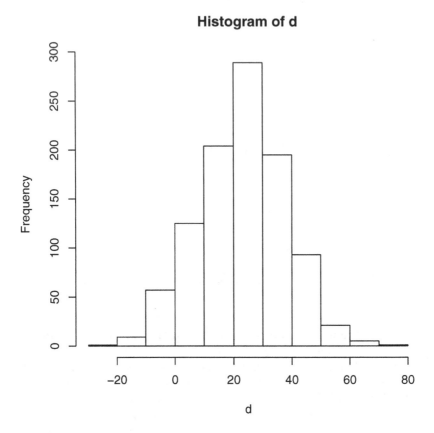

Fig. 3.14 The distribution of the difference of sample means of two samples.

$$z = \frac{\bar{x} - \mu_0}{\sigma/\sqrt{n}} = \frac{\text{sample mean} - \text{pop. mean}}{\text{sd of sampling distribution}} \tag{3.26}$$

If we replace $\bar{x}$ with d, and the new standard deviation from the two populations' standard deviations, we are ready to work out the answer:

$$z = \frac{(\bar{x}_1 - \bar{x}_2) - (\mu_1 - \mu_2)}{\sqrt{\frac{\sigma_1^2}{n_1} + \frac{\sigma_2^2}{n_2}}} \tag{3.27}$$

$$= \frac{17 - 0}{14.702} \tag{3.28}$$

$$= 1.1563 \tag{3.29}$$

Using exactly the same logic as previously, because we don't know the population parameters in realistic settings, we replace the σ's with the sample standard deviations to get the t-statistic:

$$t = \frac{(\bar{x}_1 - \bar{x}_2) - (\mu_1 - \mu_2)}{\sqrt{\frac{s_1^2}{n_1} + \frac{s_2^2}{n_2}}} \tag{3.30}$$

This is the TWO-SAMPLE t-STATISTIC.

So far so good, but we want to now translate this into a p-value, for which we need the appropriate t-curve. The problem we face here is that the degrees of freedom needed for the correct t-distribution are not obvious. The t-distribution assumes that only one s replaces a single σ; but we have two of these. If $\sigma_1 = \sigma_2$, we could just take a *weighted average* of the two sample SDs s_1 and s_2.

In our case the correct t-distribution has $n_1 - 1 + n_2 - 1$ degrees of freedom (the sum of the degrees of freedom of the two sample variances; see Rice, 1995, 422 for a formal proof).

In real life we don't know whether $\sigma_1 = \sigma_2$. One response would be to err on the side of caution, and simply use degrees of freedom corresponding to the smaller sample size. Recall that smaller degrees of freedom reflect greater uncertainty, so the estimate we get from this simple approach will be a conservative one.

However, in a more sophisticated approach, something called Welch's correction corrects for possibly unequal variances in the t-curve. R does this correction for you if you specify that the variances are to be assumed to be unequal (var.equal=FALSE).

```
> t.test.result <- t.test(sample1, sample2, mu = 0,
        alternative = "two.sided", conf.level = 0.95,
        var.equal = FALSE)
```

If you print out the contents of t.test.result, you will see detailed output. For our current discussion it is sufficient to note that the t-value is 1.35, the degrees of freedom are 12.1 (a value somewhere between the two sample sizes), and the p-value is 0.2. Recall that every time you run the t-test with newly sampled data (you should try this), your results will be slightly different; so do not be surprised if you occasionally fail to find a significant difference between the two groups even though you already know that in reality there is such a difference. We turn to this issue in the next chapter.

Problems

3.1. Choose one answer: 95% confidence intervals describe:

a. The range of individual scores
b. Plausible values for the population mean
c. Plausible values for the sample mean
d. The range of scores within one standard deviation

3.2. A 95% confidence interval has a ?% chance of describing the sample mean:

a. 95%
b. 100%

3.3. For the same data, a 90% CI will be wider than a 95% CI.

a. True
b. False

3.4. True or False?

a. The p-value is the probability of the null hypothesis being true.
b. The p-value is the probability that the result occurred by chance.

Chapter 4
Power

4.1 Hypothesis Testing Revisited

Let's assume we do an experiment, compute the t-value and p-value for our sample, and based on these values, reject the null hypothesis. As we mentioned in the previous chapter, and as you can prove to yourself through simulated replication of experiments, due to the very nature of random sampling it is always *possible* to stumble on a 'rogue sample', one whose statistic happens to be far from the population parameter. In this case it would, in fact, be an error to reject the hypothesis, though we wouldn't know it. The technical name for this is a TYPE I ERROR: the null hypothesis is true, but our sample leads us to reject it.

The converse may also happen. Suppose the null hypothesis is indeed false—there is a real difference between two population means, for example—but the samples we take happen to be so close to each other that this difference is not detectable. Here, the null hypothesis is false, but we fail to reject it based on our sample. Again, we have been misled by the sampling process: this is known as a TYPE II ERROR.

In the first case, we would think our experiment had succeeded, publish our result, and move on, unaware of our mistake. Can we protect against this? In the second case, we don't get a significant difference, it appears our experiment has failed. Is there some way to—if not prevent this—at least minimize the risk?

This aspect of statistical inference is known as POWER ANALYSIS and is of fundamental importance in experimental design. It turns out that a couple of the obvious things we *might* do to improve our experiments have unpleasant implications. For example, we might think that making the probability threshold more stringent—0.01 instead of 0.05, for example—will minimize the chance of error. This is not so. The reason is that there is an intimate relationship between the two types of error, and a technique that simply aims

S. Vasishth, M. Broe, *The Foundations of Statistics: A Simulation-based Approach*, DOI 10.1007/978-3-642-16313-5_4,
© Springer-Verlag Berlin Heidelberg 2011

to minimize one kind can unwittingly increase the chance of the other. This chapter uses simulation to explore this interaction.

4.2 Type I and Type II Errors

We fix some conventions first. Let: R = 'Reject the null hypothesis H_0'; $\neg R$ = 'Fail to reject the null hypothesis H_0.'

The decision R or $\neg R$ is based on the sample. Keep in mind that when we do an experiment we don't know whether the null hypothesis is true or not.

The first step in attempting to minimize error is to have some way to measure it. It turns out we can use probability for this as well: we will use conditional probabilities of the following kind: Let $P(R \mid H_0)$ = 'Probability of rejecting the null hypothesis *conditional on the assumption that the null hypothesis is in fact true.*'

Let's work through the logical possibilities that could hold: the null hypothesis could be in fact true or in fact false (but we don't know which), and in addition our decision could be to accept or reject the null hypothesis (see Table 4.1). In only two of these four possible cases do we make the right decision. In the table, think of α as the threshold probability we have been using all along, 0.05.

Table 4.1 The logical possibilities given the two possible situations: null hypothesis true (H_0) or false ($\neg H_0$).

Reality:	H_0	$\neg H_0$
Decision from sample is 'reject':	$P(R \mid H_0) = \alpha$ **Type I error**	$P(R \mid \neg H_0) = 1 - \beta$ **Power**
Decision from sample is 'accept':	$P(\neg R \mid H_0) = 1 - \alpha$	$P(\neg R \mid \neg H_0) = \beta$ **Type II error**

As shown in Table 4.1, the probability of a Type I error $P(R \mid H_0)$ is in fact α, conventionally set at 0.05. We will see why this is so shortly. But it immediately follows that the probability of the logical complement $P(\neg R \mid H_0)$ is $1 - \alpha$. We define the probability of a Type II error $P(\neg R \mid \neg H_0)$ to be β (more on this below), but it immediately follows that the probability of the logical complement $P(\neg R \mid \neg H_0) = 1 - \beta$. We call this probability POWER. Thus, if we want to decrease the chance of a Type II error, we need to increase the power of the statistical test.

Let's do some simulations to get a better understanding of these various definitions. We focus on the case where the null hypothesis is in fact false: there is a real difference between population means.

Assume a population with mean $\mu_1 = 60$, $\sigma_1 = 1$, and another with mean $\mu_2 = 62$, $\sigma_2 = 1$. In this case we already *know* that the null hypothesis is false. The distribution corresponding to the null hypothesis is shown in Figure 4.1. It is centered around 0, consistent with the null hypothesis that the difference between the means is 0.

We define a function for easily shading the regions of the plot we are interested in. The function below, `shadenormal2`, is a modified version of the function `shadenormal.fnc` available from the package `languageR` by Baayen, 2008 (you do not need to load the library `languageR` to use the function below).

First, we define a function that will plot Type I error intervals. This function requires that several parameters be set (our use of this function will clarify how to use these parameters):

```
> plot.type1.error <- function(x, x.min, x.max,
       qnts, mean, sd, gray.level, main, show.legend = TRUE) {
    plot(x, dnorm(x, mean, sd), type = "l", xlab = "",
        ylab = "", main = main)
    abline(h = 0)
    x1 = seq(x.min, qnorm(qnts[1]), qnts[1]/5)
    y1 = dnorm(x1, mean, sd)
    polygon(c(x1, rev(x1)), c(rep(0, length(x1)),
        rev(y1)), col = gray.level)
    x1 = seq(qnorm(qnts[2]), x.max, qnts[1]/5)
    y1 = dnorm(x1, mean, sd)
    polygon(c(x1, rev(x1)), c(rep(0, length(x1)),
        rev(y1)), col = gray.level)
    if (show.legend == TRUE) {
        legend(2, 0.3, legend = "Type I error",
            fill = gray.level, cex = 1)
    }
}
```

Next, we define a function for plotting Type I and Type II errors; this function additionally allows us to specify the mean of the null hypothesis and the population mean that the sample is drawn from (`mean.true`):

```
> plot.type1type2.error <- function(x, x.min, x.max,
       qnts, mean.null, mean.true, sd, gray1, gray2,
       main, show.legend = TRUE) {
    plot(x, dnorm(x, mean.true, sd), type = "l",
        ylab = "", xlab = "", main = main)
    lines(x, dnorm(x, mean.null, sd), col = "black")
    abline(h = 0)
    x1 = seq(qnorm(qnts[1]), x.max, qnts[1]/5)
    y1 = dnorm(x1, mean.true, sd)
```

```
    polygon(c(x1, rev(x1)), c(rep(0, length(x1)),
        rev(y1)), col = gray2)
    x1 = seq(x.min, qnorm(qnts[1]), qnts[1]/5)
    y1 = dnorm(x1, mean.null, sd)
    polygon(c(x1, rev(x1)), c(rep(0, length(x1)),
        rev(y1)), col = gray1)
    x1 = seq(qnorm(qnts[2]), x.max, qnts[1]/5)
    y1 = dnorm(x1, mean.null, sd)
    polygon(c(x1, rev(x1)), c(rep(0, length(x1)),
        rev(y1)), col = gray1)
    if (show.legend == TRUE) {
        legend(2, 0.3, legend = c("Type I error",
            "Type II error"), fill = c(gray1,
            gray2), cex = 1)
    }
}
```

The above two functions are then used within another function, **shade-normal2** (below), that plots either the Type I error probability region alone, or both Type I and Type II error probability regions. Playing with the parameter settings in this function allows us to examine the relationship between Type I and II errors.

```
> shadenormal2 <- function(plot.only.type1 = TRUE,
    alpha = 0.05, gray1 = gray(0.3), gray2 = gray(0.7),
    x.min = -6, x.max = abs(x.min), x = seq(x.min,
        x.max, 0.01), mean.null = 0, mean.true = -2,
    sd = 1, main = "", show.legend = TRUE) {
    qnt.lower <- alpha/2
    qnt.upper <- 1 - qnt.lower
    qnts <- c(qnt.lower, qnt.upper)
    if (plot.only.type1 == TRUE) {
        plot.type1.error(x, x.min, x.max, qnts,
            mean.null, sd, gray1, main, show.legend)
    }
    else {
        plot.type1type2.error(x, x.min, x.max,
            qnts, mean.null, mean.true, sd, gray1,
            gray2, main, show.legend)
    }
}
```

```
> shadenormal2(plot.only.type1 = TRUE)
```

The vertical lines in Figure 4.1 represent the 95% CI, and the shaded areas are the Type I error regions for a two-sided t-test (with probability in the two regions summing to $\alpha = 0.05$). A sample mean from one sample taken from

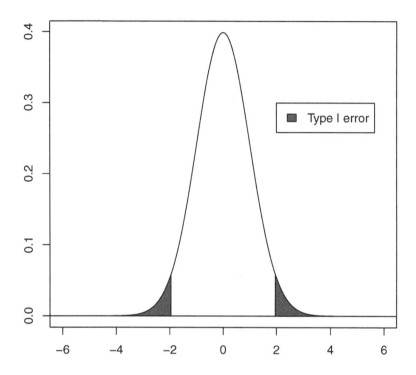

Fig. 4.1 The distribution corresponding to the null hypothesis, along with rejection regions (the Type I error probability region α).

a population with mean zero could possibly lie in this region (although it's unlikely, given the shape of the distribution), and based on that one sample, we would incorrectly decide that the null hypothesis is false when it is actually true.

In the present example we *know* there is a difference of -2 between the population means. Let's plot the *actual* (as opposed to hypothetical) sampling distribution of mean differences corresponding to this state of the world.

```
> shadenormal2(plot.only.type1 = TRUE)
> xvals <- seq(-6, 6, 0.1)
> lines(xvals, dnorm(xvals, mean = -2, sd = 1),
        lwd = 2)
```

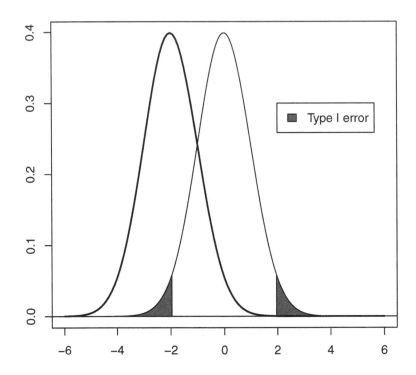

Fig. 4.2 The distribution corresponding to the null hypothesis and the distribution corresponding to the true population scores.

Figure 4.2 shows the distribution corresponding to the null hypothesis overlaid with the *actual* distribution, which we *know* is centered around −2. The vertical lines are again the 95% CI, assuming the null hypothesis is true.

Now let's shade in the region that corresponds to Type II error; see Figure 4.3. Notice that the values in this region lie *within* the 95% CI of the null hypothesis. To take a specific example, given that the population means really differ by −2, if in our particular sample the difference happened to be −1, we would fail to reject H_0 even though it is false. This is true for any value in this Type II error range.

```
> shadenormal2(plot.only.type1 = FALSE)
```

Some important insights emerge from Figure 4.3. First, if the true difference between the means had been not −2 but −4 (i.e., the EFFECT SIZE

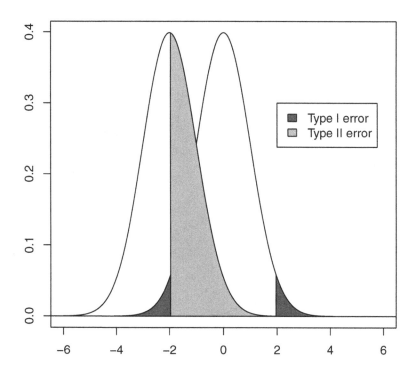

Fig. 4.3 The distribution corresponding to the true population scores along with the confidence intervals from the distribution corresponding to the null hypothesis.

had been greater), then the Type II error probability (β) will go down, and therefore power $(1 - \beta)$ will go up. Let's confirm this visually (Figure 4.4).

```
> shadenormal2(plot.only.type1 = FALSE, mean.true = -4)
```

The second insight is that if we reduce α, we also increase Type II error probability, which reduces power. Suppose α were 0.01; then the Type II error region would be as in Figure 4.5.

```
> shadenormal2(plot.only.type1 = FALSE, alpha = 0.01,
        main = "alpha=.01")
```

The third insight is that as we increase sample size, the 95% confidence intervals become tighter. This decreases Type II error probability, and therefore increases power, as shown in Figure 4.6.

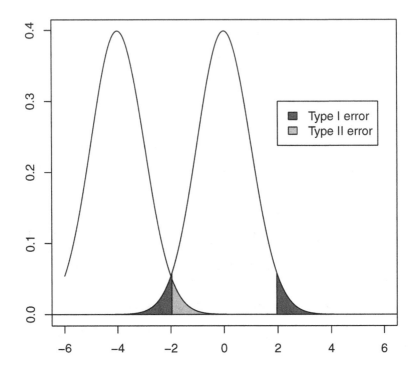

Fig. 4.4 When the true difference, i.e., the effect size, increases from -2 to -4, Type II error probability decreases, and therefore power increases. Compare with Figure 4.3.

```
> shadenormal2(plot.only.type1 = FALSE, sd = 0.75,
        main = "Larger sample size")
```

To summarize, the best situation is when we have relatively high power (low Type II error probability) and low Type I error probability (α). By convention, we keep α at 0.05. We usually do not want to change that: lowering α is costly in the sense that it reduces power as well, as we just saw. What we do want to ensure is that power is reasonably high; after all, why would you want to do an experiment where you have only 50% power or less? That would mean that you have an a priori chance of finding a true effect (i.e., an effect that is actually present in nature) only 50% of the time or less. As we just saw, we can increase power by increasing sample size, and by increasing the sensitivity of our experimental design so that we have larger effect sizes.

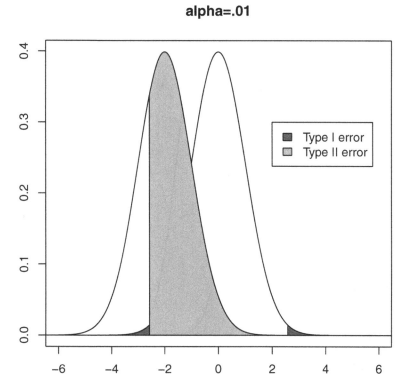

Fig. 4.5 When we decrease α from 0.05 to 0.01, Type II error probability increases, and therefore power decreases (compare Figure 4.3).

Researchers in psycholinguistics and other areas often do experiments with low power (for logistical or other reasons); it is not unheard of to publish reading studies (eyetracking or self-paced reading, etc.) or event-related potentials studies with 12–20 participants. This is not a serious problem if we succeed in getting the significant result that was predicted when the study was run. However, when we get a null (nonsignificant) result, it would be a mistake to conclude that no true effect exists (i.e., it would be a mistake to argue for the null hypothesis). If power is low, the chance of missing an effect that is actually present is high, so we should avoid concluding anything from a null result in this situation.

We would like to make four observations here:

1. At least in areas such as psycholinguistics, the null hypothesis is, strictly speaking, usually always false: When you compare reading times or any

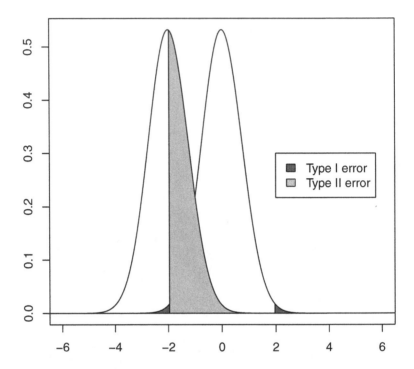

Fig. 4.6 Increasing sample size will tighten 95% confidence intervals, decreasing Type II error probability, which increases power (compare with Figure 4.3).

other dependent variable in two conditions, the a priori chance that the two means to be compared are *exactly* identical is low. The interesting question therefore is not whether the null hypothesis is false, but by how much (the effect size), and the sign (positive or negative) of the difference between the means being compared.

2. One can in principle nearly always get a statistically significant effect given a large enough sample size; the question is whether the effect is large enough to be theoretically important and whether the difference between means has the expected sign.

3. Especially in areas like psycholinguistics, replication is not given the importance it deserves. Note that we run a 5% risk of declaring a significant difference when in fact there is none (or effectively none, see point 1 above). Replication is a reliable way to convince oneself that an effect

is truly present. High power, reasonably large effect sizes, and actually replicated results should be your goal in experimental science.

4. Many researchers also believe that the lower the p-value, the lower the probability that the null hypothesis is true. However, as discussed in chapter 3, this is a misunderstanding that stems from a failure to attend to the conditional probabilities involved.

It follows from the above discussion that if you have a relatively narrow CI, and a nonsignificant result ($p > .05$), you have relatively high power and a relatively low probability of making a Type II error (accepting the null hypothesis as true when it is in fact false). This is particularly important for interpreting null results (results where the p-value is greater than 0.05). Hoenig and Heisey (2001) suggest a heuristic: if you have a narrow CI, and a nonsignificant result, you have some justification for concluding that the null hypothesis may in fact be effectively true. Conversely, if you have a wide CI and a nonsignificant result the result really is inconclusive.

The above heuristic seems a little vague; how do we define 'narrow CI'? If the goal of the experiment really is to argue for the null hypothesis, one solution is equivalence testing. The basic idea is to reverse the burden of proof. The null hypothesis becomes the alternative hypothesis and the alternative the null. What you have to determine is what you would accept as an effectively null result. For example, if the difference between two means in a reading time study is actually 5 ms, you might consider this small difference to effectively mean that the null hypothesis is essentially true.

Let us look at this approach next.

4.3 Equivalence Testing

There are two techniques that can be used for arguing in favor of a null hypothesis: TOST (Two One-Sample t-Tests); and the confidence intervals approach.

4.3.1 Equivalence Testing Example

Let's look at equivalence testing using a concrete example. This example is taken from Stegner, Bostrom, and Greenfield (1996).

We have data on two kinds of case management randomly applied to 201 seriously mentally disordered patients: (i) Traditional Case Management or TCM (Control) (ii) TCM plus trained service coordinators (Treatment). Treatment is costlier than Control, so if they are not significantly different the extra cost is money wasted. The dependent measure is Brief Psychiatric

Rating Scale (for our purposes, it doesn't matter how this is calculated). The data summary is in Table 4.2 (Some patients' data were not available).

Table 4.2 Data from Stegner et al. 1996.

Group	n	Mean	SD
Control	64	1.5679	0.4285
Treatment	70	1.6764	0.4748
Total	134	1.6246	0.4533 (pooled)

Let $\bar{x}_C$ be the mean for Controls, and $\bar{x}_T$ the mean for Treatment, and let the pooled standard deviation be s_{pooled}. Specifically, $\bar{x}_C = 1.5679$, $\bar{x}_T = 1.6764$, and $s_{pooled} = 0.4533$ Therefore, the difference between the two means d is: $d = \bar{x}_T - \bar{x}_C = 1.6764 - 1.5679 = 0.1085$.

Here, the research goal is to find out if the treatment is effective or not; if it is not, the difference between the means should be 'essentially' equivalent. In order to formally specify what is meant by 'essentially' equivalent, we can specify an equivalence threshold Θ; if d lies within this threshold we accept the alternative hypothesis (effectively no difference–the traditional null hypothesis). Suppose previous experience in the field suggests that a difference of 20% or less with respect to the Control's mean can be considered to be equivalent. $\Theta = .2 \times 1.5679 = 0.3136$. There should be some independent, prior criterion for deciding what Θ will be.

4.3.2 TOST Approach to the Stegner et al. Example

Since $\Theta = 0.3136$, we can define two limits around 0 that constitute the equivalence threshold: $\Theta_L = -0.3136, \Theta_U = 0.3136$. If d lies within this region we reject the hypothesis that the two means are different. Thus, our null and alternative hypotheses are:

$$H_0 : d \leq \Theta_L \text{ or } d \geq \Theta_U \tag{4.1}$$

$$H_a : \Theta_L < d < \Theta_U \tag{4.2}$$

We then compute two one-sided t-tests to evaluate each of the two parts of the null hypothesis:

$$t = \frac{d - \Theta}{SE} = \frac{d - \Theta}{s_{pooled}/\sqrt{(1/n_1 + 1/n_2)}} = -2.616 \tag{4.3}$$

$$t = \frac{d + \Theta}{SE} = \frac{d + \Theta}{s_{\text{pooled}}/\sqrt{(1/n_1 + 1/n_2)}} = 5.384 \qquad (4.4)$$

Since the critical t-value is $t(134 - 2) = 1.6565$ (the R command qt(.95, df = 132) yields this t-value), both parts of the null hypothesis are rejected. It follows that the difference between the two population means is no greater than Θ; the extra cost is unjustified.

Summary of calculations:

1. Define an equivalence threshold Θ.
2. Compute two one-way t-tests:

$$t_{d \leq \Theta_L} = \frac{d - \Theta}{s_{\text{pooled}}/\sqrt{(1/n_1 + 1/n_2)}} \qquad (4.5)$$

$$t_{d \geq \Theta_U} = \frac{d + \Theta}{s_{\text{pooled}}/\sqrt{(1/n_1 + 1/n_2)}} \qquad (4.6)$$

3. Compute critical t-value t_{crit} (the 95% CI cutoff points). In R this is done as follows: qt(.95,DF), where DF=$n_1 + n_2 - 2$; n_1 and n_2 are the sample sizes of the two groups.
4. If $t_{d \leq \Theta_L} < -t_{\text{crit}}$ and if $t_{d \geq \Theta_U} > t_{\text{crit}}$, we can reject the null hypothesis.

It's easy to write a function that does the above calculation for us:

```
> TOST <- function(mean1, mean2, theta, n1, n2,
      sigma) {
      d <- (mean2 - mean1)
      t1 <- (d - theta)/(sigma * (sqrt((1/n1) +
          (1/n2))))
      t2 <- (d + theta)/(sigma * (sqrt((1/n1) +
          (1/n2))))
      tcrit <- qt(0.95, (n1 + n2 - 2))
      if ((t1 < -tcrit) && (t2 > tcrit)) {
          print(t1)
          print(t2)
          print(tcrit)
          print(c("Equivalent"))
      }
      else {
          print(c("Failed to show equivalence"))
      }
  }
```

4.3.3 Equivalence Testing Example: CIs Approach

Schuirmann (1987) showed that TOST is operationally equivalent to determining whether $100(1\text{-}2\alpha)\%$ CIs fall within the range $-\Theta \cdots +\Theta$. Recall that $t_{\text{crit}} = 1.6565$. We can now compute the confidence intervals:

$$CI = d \pm 1.6565 \times SE \qquad (4.7)$$

$$= d \pm 1.6565 \times \left(\frac{\sigma}{\sqrt{(1/n_1 + 1/n_2)}}\right) \qquad (4.8)$$

$$= 0.1085 \pm 1.6565 \times \left(\frac{0.4533}{\sqrt{(1/64 + 1/70)}}\right) \qquad (4.9)$$

$$= 0.1085 \pm 0.1299 \qquad (4.10)$$

Since $(-0.0214, 0.2384)$ lies within the range $(-0.3136, +0.3136)$ we can declare equivalence. Recall now the heuristic we gave earlier: narrow CIs, accept null hypothesis; wide CIs, inconclusive.

4.4 Observed Power and Null Results

Some journals and organizations ask the researcher to compute OBSERVED POWER (defined below) if you get a null result. The logic is: if you got a null result ($p > .05$) and the observed power based on the sample is high ($> .80$), then you can safely accept the null result as true. After all, $P(R \mid \neg H_0) > .80$, implying the chance of a Type II error is low. So if the rejection of the hypothesis is not due to the hypothesis being false, it must be a case of rejection due to inability of the sample to detect the difference.

The problem with this is that the p-value and observed power are inversely related. Observed power provides no new information after the p-value is known. Let's convince ourselves this is true with a simple example.

Take the earlier example of a population with mean $\mu_1 = 60$, $\sigma_1 = 1$, and another with $\mu_2 = 62$, $\sigma_2 = 1$. If we compare the means from two samples, one taken from each population, and our null hypothesis is that the two samples come from the same population, we already know in this case that the null hypothesis is false (the difference between the means is -2). But suppose we didn't know this, and we got a sample mean difference of -1.9 and some p-value $p > 0.05$. We can compute observed power using this *observed* difference (cf. the *actual* difference -2 used earlier).

But if our sample mean difference had been -1 and the associated p-value p' had therefore been greater than p, we could also have computed observed power. This can be illustrated as follows (the resulting figure is shown in Figure 4.7).

```
> multiplot(1, 2)
> shadenormal2(plot.only.type1 = FALSE, mean.true = -1.9,
        main = "Lower p,\n higher observed power",
        show.legend = FALSE)
> shadenormal2(plot.only.type1 = FALSE, mean.true = -1,
        main = "Higher p, \n lower observed power",
        show.legend = FALSE)
```

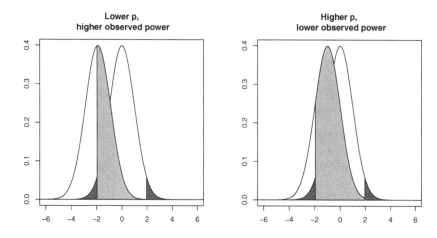

Fig. 4.7 Observed power and p-values are inversely related, so once we know the p-value, the observed power provides no new information.

Figure 4.7 shows that the area under the curve outside the vertical lines (power) decreases as p-values go up (as the difference in the sample means comes closer to zero).[1] For this reason, computing observed power does not provide any new information: if the p-value is high, we already know the observed power is low, there is nothing gained by computing it.

Another suggestion one sometimes hears in experimental science is to keep increasing sample size n until you get a significant difference. Recall that $\sqrt{n}$ is inversely related to SE, which is used to compute 95% CIs. So, by increasing sample size, you are narrowing the CIs, and so eventually (in the limit) you are guaranteed to get some difference between the means (although whether it has the sign—positive or negative—that you want or expect is another question).

---

[1] In the code for this figure, we abuse the notation for the parameter mean.true a little bit to illustrate the point; here, it stands for the *observed* mean from a sample, not the true population mean.

A much better approach is to do power analysis *before* running the experiment. Here, based on one's prior knowledge of a phenomenon (e.g., from previous work in the area, pilot studies, model predictions, or intuition-based estimates), one decides on an effect size one expects, and the power calculation delivers the number of participants one needs for a power level specified by the user. R has a library, `pwr`, written by Stephane Champely, that computes power for various test statistics following the proposals set forth in (Cohen, 1988). See the documentation for that package for details.

Problem

4.1. Suppose you carry out a between-participants experiment where you have a control and treatment group, say 20 participants in each sample. You carry out a two-sample t-test and find that the result is t(18)=2.7, p<0.01. Which of the statements below are true?

1. You have absolutely disproved the null hypothesis.
2. The probability of the null hypothesis being true is 0.01.
3. You have absolutely proved that there is a difference between the two means.
4. You have a reliable experimental finding in the sense that if you were to repeat the experiment 100 times, in 99% of the cases you would get a significant result.

This problem is adapted from a book by Oakes (1987, 79) (we highly recommend reading this book).

Chapter 5
Analysis of Variance (ANOVA)

5.1 Comparing Three Populations

Rietveld and van Hout (2005) provide this fictional example involving three second-language vocabulary learning methods (I, II, III), with three different groups of participants assigned to each method.[1] The relative effectiveness of the learning methods is evaluated on some scale by scoring the increase in vocabulary after using the method. We draw a sample from each group, and compute the mean scores

Table 5.1 Fictional dataset from Rietveld and van Hout (2005).

	Group I	Group II	Group III
	9	10	1
	1	2	5
	2	6	0
$\bar{x}$	4	6	2

Consider our research question: is any one of the learning methods better than the others? Can we employ the t-test technique in this situation? We effectively have three populations here, not two. We might reason as follows: do a two-sample t-test on Group I vs. II, I vs. III, II vs. III; if *at least one* of the three null hypotheses can be rejected, we can safely reject the main research hypothesis that the means of the three groups are the same.

We can represent the three null hypotheses as follows:

---

[1] The initial part of this chapter follows the presentation of analysis of variance (ANOVA) by Rietveld and van Hout (2005); the reader looking for more details on ANOVA is encouraged to consult that book.

S. Vasishth, M. Broe, *The Foundations of Statistics: A Simulation-based Approach*, DOI 10.1007/978-3-642-16313-5_5,
© Springer-Verlag Berlin Heidelberg 2011

$$A \rightarrow H_{0_{I,II}} : \mu_I = \mu_{II} \tag{5.1}$$

$$B \rightarrow H_{0_{I,III}} : \mu_I = \mu_{III} \tag{5.2}$$

$$C \rightarrow H_{0_{II,III}} : \mu_{II} = \mu_{III} \tag{5.3}$$

The eight logically possible outcomes associated with the three t-tests are listed in Table 5.2 (let $\neg$ X mean 'reject X'):

Table 5.2 All possible outcomes of pairwise comparisons.

1st	2nd	3rd	4th	5th	6th	7th	8th
A	$\neg$A	A	A	$\neg$A	A	$\neg$A	$\neg$A
B	B	$\neg$B	B	$\neg$B	$\neg$B	B	$\neg$B
C	C	C	$\neg$C	C	$\neg$C	$\neg$C	$\neg$C

Let the probability of rejecting the null hypothesis when it is true, i.e., the α level, be 0.05. The probability of rejecting at least one null hypothesis is the sum of the probabilities of each of the mutually exclusive events in the 2nd to 8th columns. Hence, the probability of rejecting at least one null hypothesis, assuming that it is actually true, is no longer 0.05 but:

$$3 \times 0.05 \times (0.95)^2 + 3 \times 0.95 \times (0.05)^2 + (0.05)^3 = 0.142625 \tag{5.4}$$

So now our α level for rejecting the overall null hypothesis 'all methods are equal' is no longer 0.05, and sets a lower standard for rejection than we would like. Performing a set of pairwise t-tests is not a good solution to our problem. We need a more holistic approach.

Imagine now that we replicate our experiment and add a second set of samples for comparison:

Table 5.3 Fictional data from Rietveld and van Hout, (2005).

	Group I	Group II	Group III	‖	Group I	Group II	Group III
	9	10	1	‖	3	7	1
	1	2	5	‖	4	6	2
	2	6	0	‖	5	5	3
$\bar{x}$	4	6	2	‖	4	6	2

Group II seems to be doing consistently better across both experiments. However, an important difference between the first and second sample is that in the first there is a lot more variance within groups. The variation in the mean scores (between-group variation) in the first sample could just be due to within-group variation. In the second sample, there's a lot less within-group variation, but the between-group variation is just the same—maybe Group II really is doing significantly better. What we just did was (informally) analyze the variance between and within groups—hence the name of this procedure: ANALYSIS OF VARIANCE or ANOVA.

5.2 ANOVA

In the early days of astronomical observation, it was noticed that increasing precision with measurement came with a price: different observations of the same event rarely agreed exactly. It was initially thought that pooling the observations in any way would also pool the errors, making the situation worse. Gauss noticed that the observational error had a particular distribution: there were more observations close to the truth than not, and errors overshot and undershot with equal probability. The errors in fact have a normal (or 'gaussian') distribution. Thus if we average the observations, the errors tend to cancel themselves out! Notice we can think of any particular observation then as 'containing' the true value as well as an error term.

We can think of a sample mean in the same way. Any sample mean can be thought of as 'containing' the true population mean plus an error term. Samples from the same population then contain the same true mean, plus differing error terms:

$$\bar{x}_1 = \mu + \varepsilon_1 \tag{5.5}$$
$$\bar{x}_2 = \mu + \varepsilon_2 \tag{5.6}$$
$$\bar{x}_3 = \mu + \varepsilon_3 \tag{5.7}$$

This way of thinking about a sample value is reminiscent of the numerator in the z-score: the error is simply the distance from the mean. We can see this in the following simulation (Figure 5.1):

```
> errors <- rep(NA, 1000)
> for (i in 1:1000) {
      sample <- rnorm(10, mean = 60, sd = 1)
      errors[i] <- mean(sample) - 60
  }
> hist(errors)
```

In what follows, we will use this fact to build statistical models.

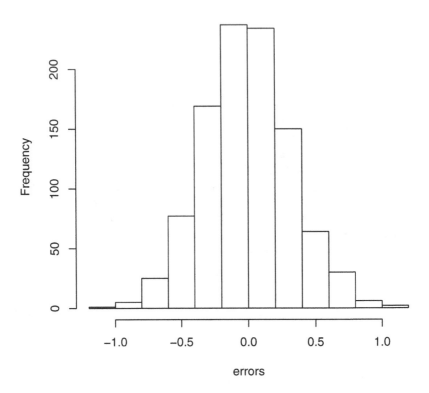

Fig. 5.1 Errors tend to cancel out.

5.2.1 Statistical Models

Characterizing a sample mean as an error about a population mean is perhaps the simplest possible example of building a STATISTICAL MODEL:

$$\bar{x}_j = \mu + \varepsilon_j \tag{5.8}$$

It's a powerful idea because it allows us to *compare* statistical models and decide which one better characterizes the data. We could consider different possibilities for μ, for example, and see which one minimizes the error, i.e., which provides the best 'fit' to the data. In subsequent chapters we'll be looking at just how powerful this idea is. The way we use it here however, is to let one model represent the null hypothesis (no difference between means), and another the alternative hypothesis (systematic difference between means).

Following the presentation of Rietveld and van Hout (2005), let's gradually build a model for the current example, and systematically break apart the values in each cell into model parameters and error terms.

We start with an idealized dataset first (Table 5.4).

Table 5.4 The fictional dataset with identical scores in each cell.

	Group I	Group II	Group III
	$x_{1,1} = 4$	$x_{1,2} = 4$	$x_{1,3} = 4$
	$x_{2,1} = 4$	$x_{2,2} = 4$	$x_{2,3} = 4$
	$x_{3,1} = 4$	$x_{3,2} = 4$	$x_{2,3} = 4$
$\bar{x} = 4$	$\bar{x}_1 = 4$	$\bar{x}_2 = 4$	$\bar{x}_3 = 4$

Here there is no variation of any kind, so the model for i subjects, $i = 1, \ldots, n$, and j groups, $j = 1, \ldots, m$, is:

$$x_{i,j} = \mu \tag{5.9}$$

(Pay particular attention to the subscripts for group on the one hand, and individual subjects on the other. The order $x_{subject,group}$ is chosen to be consistent with R's conventions). Here, μ is 4.

Now look at this slightly different data (Table 5.5), the only change is that it has some variation between groups:

Table 5.5 Fictional data with different mean scores for each group.

	Group I	Group II	Group III
	$x_{1,1} = 4$	$x_{1,2} = 6$	$x_{1,3} = 2$
	$x_{2,1} = 4$	$x_{2,2} = 6$	$x_{2,3} = 2$
	$x_{3,1} = 4$	$x_{3,2} = 6$	$x_{2,3} = 2$
$\bar{x} = 4$	$\bar{x}_1 = 4$	$\bar{x}_2 = 6$	$\bar{x}_3 = 2$

Note that the GRAND MEAN $\bar{x}$ (the mean of means) is still 4. The model now changes to:

$$x_{i,j} = \mu + \alpha_j \tag{5.10}$$

μ is still 4, but $\alpha_1 = 0, \alpha_2 = 2, \alpha_3 = -2$, see Table 5.6.

Think about what α_j means: it models the variation between groups. Now, more realistically, scores also show variation *within* a group of subjects—individual subjects differ. For each subject i in each group j, we can represent this within-subject variation as a further term in the model, an error component ε_{ij}, around the group mean (equation 5.11):

Table 5.6 Each group has different mean scores, which can be characterized as a deviation from the grand mean.

	Group I	Group II	Group III
	$x_{1,1} = 4 + \mathbf{0}$	$x_{1,2} = 4 + \mathbf{2}$	$x_{1,3} = 4 - \mathbf{2}$
	$x_{2,1} = 4 + \mathbf{0}$	$x_{2,2} = 4 + \mathbf{2}$	$x_{2,3} = 4 - \mathbf{2}$
	$x_{3,1} = 4 + \mathbf{0}$	$x_{3,2} = 4 + \mathbf{2}$	$x_{2,3} = 4 - \mathbf{2}$
$\bar{x} = 4$ $\bar{x}_1 = 6$		$\bar{x}_2 = 6$	$\bar{x}_3 = 2$

$$x_{i,j} = \mu + \alpha_j + \varepsilon_{ij} \qquad (5.11)$$

We can augment tables showing both datasets to include this error component, Tables 5.7 and 5.8:

Table 5.7 Scores decomposed into the grand mean, the effects of each group, and the error component (first 'experiment').

	Group I	Group II	Group III
	$4 + 0 + 5 = 9$	$4 + 2 + 4 = 10$	$4 - 2 - 1 = 1$
	$4 + 0 - 3 = 1$	$4 + 2 - 4 = 2$	$4 - 2 + 3 = 5$
	$4 + 0 - 2 = 2$	$4 + 2 + 0 = 6$	$4 - 2 - 2 = 0$
$\bar{x} = 4$ $x_1 = 4$		$x_2 = 6$	$x_3 = 2$

Table 5.8 Scores decomposed into the grand mean, the effects of each group, and the error component (second 'experiment').

	Group I	Group II	Group III
	$4 + 0 - 1 = 3$	$4 + 2 + 1 = 7$	$4 - 2 - 1 = 1$
	$4 + 0 + 0 = 4$	$4 + 2 + 0 = 6$	$4 - 2 + 0 = 2$
	$4 + 0 + 1 = 5$	$4 + 2 - 1 = 5$	$4 - 2 + 1 = 3$
$\bar{x} = 4$ $x_1 = 4$		$x_2 = 6$	$x_3 = 2$

In this fabricated example, the parameter values are known to be $\mu = 4$, and $\alpha_1 = 0, \alpha_2 = 2, \alpha_3 = -2$: in real life we would have to estimate them. Our task here is simpler however. We want to use the model to construct a null hypothesis. If there is no systematic variation between the population means in each group, then

$$\alpha_1 = \alpha_2 = \alpha_3 = 0 \qquad (5.12)$$

In this case the variation observed between sample means is *due only to error variation within the samples*. If an effect α_j is present, the variation between groups increases because of the systematic differences between groups: the between-group variation is due to error variation plus variation due to α_j. So the null hypothesis becomes:

$$H_0 : x_{ij} = \mu + \varepsilon_{ij} \tag{5.13}$$

The alternative hypothesis is that $\alpha \neq 0$, and we can write this as:

$$H_a : x_{ij} = \mu + \alpha_j + \varepsilon_{ij} \tag{5.14}$$

Where do we go from here? What kind of evidence counts against the null hypothesis?

5.2.2 Variance of Sample Means as a Possible Statistic

In order to do statistical inference for this situation, we need to know what the world would look like if the null hypothesis were true. Recall the situation for two populations that we discussed in chapter 3 (page 76): we chose a statistic—the difference between the sample means d—and found what its sampling distribution would look like if there really were no difference between the population means:

$$H_0 : \mu_1 - \mu_2 = \delta = 0. \tag{5.15}$$

We saw that the relevant sampling distribution was a normal distribution whose mean is 0. We then measure how far from this mean our actual sample difference lies (its t-value), in order to test the hypothesis.

What statistic could we use in the present case? An obvious possibility is the *variance of the sample means*. It has some nice-looking properties: it gets bigger as the three or more sample means get spread further apart, and is equal to 0 if all the means are the same—all apparently similar to the difference of the means d, which is an unbiased estimator of δ. For example, if we had three means, 60, 62, 64, then the variance of the three means would be 4; if the three means were, e.g., 50, 60, 70, then it would be 100; and if the three means were all 60, then it would be 0:

```
> var(c(60, 62, 64))

[1] 4

> var(c(50, 60, 70))
```

```
[1] 100

> var(c(60, 60, 60))

[1] 0
```

Using this statistic as a measure, what would the world look like (what would the sampling distribution be) if the null hypothesis were true? Let's do a simulation to find out. We sample from three populations, all of whom have the same population mean, and compute the variance of the sample means (Figure 5.2).

```
> variances <- rep(NA, 1000)
> for (i in 1:1000) {
        sample1 <- rnorm(11, mean = 60, sd = 4)
        sample2 <- rnorm(11, mean = 60, sd = 4)
        sample3 <- rnorm(11, mean = 60, sd = 4)
        variances[i] <- var(c(mean(sample1), mean(sample2),
            mean(sample3)))
   }
> mean(variances)

[1] 1.435202

> plot(density(variances), main = "", xlab = "",
        ylab = "")
```

Examine Figure 5.2 and consider the following example of a single replicate:

```
> sample1 <- rnorm(11, mean = 60, sd = 4)
> sample2 <- rnorm(11, mean = 60, sd = 4)
> sample3 <- rnorm(11, mean = 60, sd = 4)
> c(mean(sample1), mean(sample2), mean(sample3))
[1] 61.81673 57.94935 60.67934
> var(c(mean(sample1), mean(sample2), mean(sample3)))
[1] 3.950507
```

The problem here is the sampling variation (the error) apparent in the fluctuating means does not 'cancel out' when we measure their variance: the variance will always be some positive number unless the sample means all happen to be identical, which is highly improbable in itself. Sometimes the variance will be quite high (as is clear from the plot) and these values can never be 'balanced' by even the lowest variance (0). Hence the mean of this sampling distribution is much greater than 0.

As the sample size goes up, the sample means will be tighter, and the variance will go down, but it will always be positive and skewed right, and thus the mean of this sampling distribution will always overestimate the true parameter. Considerations such at this prompt us to search for a different statistic to characterize the null hypothesis in this case.

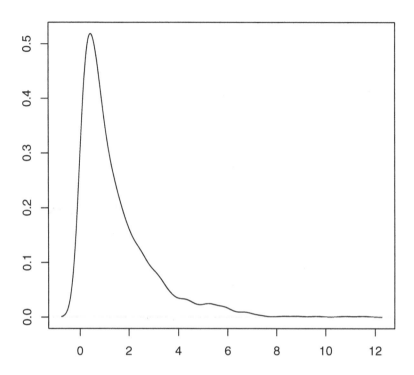

Fig. 5.2 The sampling distribution of the sample mean variance.

5.2.3 Analyzing the Variance

In order to identify an appropriate statistic, we need to do some basic algebraic manipulation.

Notice that the following equality holds (recall that i ranges across participants within a group, and j ranges across groups):

$$x_{ij} = x_{ij} \tag{5.16}$$

$$x_{ij} - \bar{x} = x_{ij} - \bar{x} \tag{5.17}$$

$$= x_{ij} + (-\bar{x}_j + \bar{x}_j) - \bar{x} \tag{5.18}$$

$$= (x_{ij} - \bar{x}_j) + (\bar{x}_j - \bar{x}) \tag{5.19}$$

$$= (\bar{x}_j - \bar{x}) + (x_{ij} - \bar{x}_j) \tag{5.20}$$

So for all i, j:

$$x_{ij} - \bar{x} = (\bar{x}_j - \bar{x}) + (x_{ij} - \bar{x}_j) \tag{5.21}$$

That is, the difference between any value and the grand mean is equal to the sum of (i) the difference between that value and its group mean and (ii) the difference between its group mean and the grand mean. Note that this is not some special property of this particular case, it is true of any 3 numbers x, y, z:

$$x - z = (x - y) + (y - z) \tag{5.22}$$

and is obvious if you plot some values on a number line. Nevertheless, this almost trivial observation is the essential fact that will allow us to break apart the total variance into 'within-groups variance' (the variance of the error) and 'between-groups variance' and thus formulate an appropriate statistic.

We can see how all this works for our specific example (Table 5.9) using R's matrix capabilities.

Table 5.9 Fictional dataset from Rietveld and van Hout (2005).

	Group I	Group II	Group III
	9	10	1
	1	2	5
	2	6	0
$\bar{x}$	4	6	2

First, we prepare a matrix containing the three groups' scores:

```
> rvh <- matrix(c(9, 1, 2, 10, 2, 6, 1, 5, 0),
        3, 3)

       [,1] [,2] [,3]
[1,]    9   10    1
[2,]    1    2    5
[3,]    2    6    0
```

The grand mean is computed simply as `mean(rvh)`; the mean of the each group j, where j ranges from 1–3, as `mean(rvh[,j])`.

We first compute the difference between each value and the grand mean, which we call `total` (the reason will become apparent):

```
> total <- matrix(rep(NA, 9), 3, 3)
> for (j in 1:3) {
        for (i in 1:3) {
```

```
        total[i, j] <- rvh[i, j] - mean(rvh)
    }
}
> print(total)

    [,1] [,2] [,3]
[1,]    5    6   -3
[2,]   -3   -2    1
[3,]   -2    2   -4
```

An aside on matrix calculations:
Don't be intimidated by the nested **for**-loops: this is simply a way to systematically visit each cell in the matrix and is a very common computational technique for tables like this. You can read it as 'for every row in every column... do something.' Just focus on what gets done in each cell, buried within the inner loop. Note here that there is an easier and faster way to carry out the above computation:

```
> total <- rvh - mean(rvh)

    [,1] [,2] [,3]
[1,]    5    6   -3
[2,]   -3   -2    1
[3,]   -2    2   -4
```

For ease of exposition we will continue to use the nested **for**-loops; the reader should, as an exercise, try rewriting the matrix-based computations shown below using more efficient code.

Second, we compute the 'within' group differences, i.e., the difference between each value and its own group mean:

```
> within <- matrix(rep(NA, 9), 3, 3)
> for (j in 1:3) {
        for (i in 1:3) {
            within[i, j] <- rvh[i, j] - mean(rvh[,
                j])
        }
    }
> print(within)

    [,1] [,2] [,3]
[1,]    5    4   -1
[2,]   -3   -4    3
[3,]   -2    0   -2
```

And finally, we compute the 'between' group differences, i.e., the difference between each group mean and the grand mean:

```
> between <- matrix(rep(NA, 9), 3, 3)
> for (j in 1:3) {
        for (i in 1:3) {
                between[i, j] <- mean(rvh[, j]) - mean(rvh)
        }
    }
> print(between)

     [,1] [,2] [,3]
[1,]   0    2   -2
[2,]   0    2   -2
[3,]   0    2   -2
```

We can check that for every value, it is in fact equivalent to the sum indicated above in equation 5.2.3:

```
> total

     [,1] [,2] [,3]
[1,]    5    6   -3
[2,]   -3   -2    1
[3,]   -2    2   -4

> within + between

     [,1] [,2] [,3]
[1,]    5    6   -3
[2,]   -3   -2    1
[3,]   -2    2   -4
```

We now want to aggregate these individual differences. Let's unpack the above equation:

$$x_{11} - \bar{x} = (\bar{x}_1 - \bar{x}) + (x_{11} - \bar{x}_1) \tag{5.23}$$

$$x_{21} - \bar{x} = (\bar{x}_1 - \bar{x}) + (x_{21} - \bar{x}_1) \tag{5.24}$$

$$x_{31} - \bar{x} = (\bar{x}_1 - \bar{x}) + (x_{31} - \bar{x}_1) \tag{5.25}$$

$$x_{12} - \bar{x} = (\bar{x}_2 - \bar{x}) + (x_{12} - \bar{x}_1) \tag{5.26}$$

$$x_{22} - \bar{x} = (\bar{x}_2 - \bar{x}) + (x_{22} - \bar{x}_1) \tag{5.27}$$

$$x_{32} - \bar{x} = (\bar{x}_2 - \bar{x}) + (x_{32} - \bar{x}_1) \tag{5.28}$$

$$x_{13} - \bar{x} = (\bar{x}_3 - \bar{x}) + (x_{13} - \bar{x}_1) \tag{5.29}$$

$$x_{23} - \bar{x} = (\bar{x}_3 - \bar{x}) + (x_{23} - \bar{x}_1) \tag{5.30}$$

$$x_{33} - \bar{x} = (\bar{x}_3 - \bar{x}) + (x_{33} - \bar{x}_1) \tag{5.31}$$

The 'left column' above is all the total differences (which we have stored in `total`), the second column is all the between-group differences (stored in `between`), and the right column is all the within-group differences (stored in `within`). Due to all the equalities, it's clear that the sum of the left hand sides equals the sum of the right hand sides:

$$\sum_{j=1}^{I}\sum_{i=1}^{n_j} x_{ij} - \bar{x} = \sum_{j=1}^{I}\sum_{i=1}^{n_j}((\bar{x}_j - \bar{x}) + (x_{ij} - \bar{x}_j)) \tag{5.32}$$

This is true, but doesn't get us very far (make sure you understand why... recall the discussion of error above):

```
> sum(total)

[1] 0

> sum(between)

[1] 0

> sum(within)

[1] 0
```

Accordingly, we square the terms on both sides, and then sum:

$$\sum_{j=1}^{I}\sum_{i=1}^{n_j}(x_{ij} - \bar{x})^2 = \sum_{j=1}^{I}\sum_{i=1}^{n_j}((\bar{x}_j - \bar{x}) + (x_{ij} - \bar{x}_j))^2 \tag{5.33}$$

```
> sum(total^2)

[1] 108

> sum(between^2, within^2)

[1] 108
```

It is now fairly easy to show that we can break apart the right hand side:

$$\sum_{j=1}^{I}\sum_{i=1}^{n_j}(x_{ij} - \bar{x})^2 = \sum_{j=1}^{I}\sum_{i=1}^{n_j}(\bar{x}_j - \bar{x})^2 + \sum_{j=1}^{I}\sum_{i=1}^{n_j}(x_{ij} - \bar{x}_j)^2 \tag{5.34}$$

```
> sum(total^2)

[1] 108

> sum(between^2) + sum(within^2)

[1] 108
```

Here is the proof:

$$\sum_{j=1}^{I}\sum_{i=1}^{n_j}(x_{ij}-\bar{x})^2 = \sum_{j=1}^{I}\sum_{i=1}^{n_j}((\bar{x}_j-\bar{x})+(x_{ij}-\bar{x}_j))^2 \tag{5.35}$$

$$= \sum_{j=1}^{I}\sum_{i=1}^{n_j}((\bar{x}_j-\bar{x})^2+(x_{ij}-\bar{x}_j)^2+\underline{2(\bar{x}_j-\bar{x})(x_{ij}-\bar{x}_j)}) \tag{5.36}$$

It's enough to show that the underlined part $= 0$.

$$\sum_{j=1}^{I}\sum_{i=1}^{n_j}2(\bar{x}_j-\bar{x})(x_{ij}-\bar{x}_j) = \sum_{j=1}^{I}2(\bar{x}_j-\bar{x})\sum_{i=1}^{n_j}(x_{ij}-\bar{x}_j) \tag{5.37}$$

Notice that for any group j the following holds (and you should certainly know why!):

$$\sum_{i=1}^{n_j}(x_{ij}-\bar{x}_j) = 0. \tag{5.38}$$

We have just established that the total SUM OF SQUARES (SS-TOTAL) is the sum of the SS-BETWEEN and SS-WITHIN groups:

$$\sum_{j=1}^{I}\sum_{i=1}^{n_j}(x_{ij}-\bar{x})^2 = \sum_{j=1}^{I}\sum_{i=1}^{n_j}(\bar{x}_j-\bar{x})^2 + \sum_{j=1}^{I}\sum_{i=1}^{n_j}(x_{ij}-\bar{x}_j)^2 \tag{5.39}$$

$$SS_{\text{total}} = SS_{\text{between}} + SS_{\text{within}} \tag{5.40}$$

Now recall the definition of variance:

$$s^2 = \frac{\sum_{i=1}^{n}(x_i-\bar{x})^2}{n-1} \tag{5.41}$$

Also recall that $n-1$ is the degrees of freedom. The numerator of the variance is precisely a sum of squares:

$$\sum_{i=1}^{n}(x_i-\bar{x})^2 \tag{5.42}$$

So to get to the variances within and between each group, we simply need to divide each SS by the appropriate degrees of freedom. How do we get these degrees of freedom? The DF-total and DF-between are analogous to the case for the simple variance ($n-1$, $I-1$). The DF-within is not so obvious; however, keep in mind that DF-total = DF-between + DF-within, so that:

$$\text{DF-within} = \text{DF-total} - \text{DF-between} \tag{5.43}$$
$$= (n-1) - (I-1) \tag{5.44}$$
$$= n - I \tag{5.45}$$

The number of scores minus the number of parameters estimated (here, the number of means) gives you the degrees of freedom for each variance. The logic for this is identical to the reason why we have $n-1$ as a denominator for variance σ^2. These are shown in the equations below, where N is the total of all the sample sizes, and I is the number of groups. Note that another (in fact more general) term for a variance is the MEAN SQUARE (MS), which is the term used in ANOVA.[2]

$$MS_{\text{total}} = \frac{\sum\limits_{j=1}^{I} \sum\limits_{i=1}^{n_j} (x_{ij} - \bar{x})^2}{N-1} \tag{5.46}$$

$$MS_{\text{between}} = \frac{\sum\limits_{j=1}^{I} \sum\limits_{i=1}^{n_j} (\bar{x}_j - \bar{x})^2}{I-1} \tag{5.47}$$

$$MS_{\text{within}} = \frac{\sum\limits_{j=1}^{I} \sum\limits_{i=1}^{n_j} (x_{ij} - \bar{x}_j)^2}{N-I} \tag{5.48}$$

In our particular example we have:

```
> ms.total <- sum(total^2)/(9 - 1)

[1] 13.5

> ms.between <- sum(between^2)/(3 - 1)

[1] 12

> ms.within <- sum(within^2)/(9 - 3)

[1] 14
```

5.3 Hypothesis Testing

We now have a number of new statistics with which we can formulate and explore the null hypothesis and perform statistical inference using a sample.

---

[2] To remind you: The sum of deviations from mean is always zero, so if we know $n-1$ of the deviations, the last deviation is predictable. The mean is an average of n unrelated numbers, but s is an average of $n-1$ unrelated numbers.

Recall our null and alternative hypotheses:

$$H_0 : x_{ij} = \mu + \varepsilon_{ij} \tag{5.49}$$

$$H_a : x_{ij} = \mu + \alpha_j + \varepsilon_{ij} \tag{5.50}$$

The null hypothesis amounts to saying that there is no effect of α_j: that any between group variance we see is completely attributable to within group variance. In other words:

$$H_0 : MS_{\text{between}} = MS_{\text{within}} \tag{5.51}$$

Alternatively, we can state the null hypothesis as a ratio:

$$H_0 : \frac{MS_{\text{between}}}{MS_{\text{within}}} = 1 \tag{5.52}$$

5.3.1 MS-within, MS-between as Statistics

Let's examine the sampling distributions of these statistics when the null hypothesis is true, using simulations. We will sample from three populations with identical means $\mu = 60$, and standard deviations $\sigma = 4$, $\sigma^2 = 16$, and compute the relevant statistics each time (exactly as we did earlier).

```
> ms.withins <- rep(NA, 1000)
> ms.betweens <- rep(NA, 1000)
> fs <- rep(NA, 100)
> for (n in 1:1000) {
        sample1 <- rnorm(11, 60, 4)
        sample2 <- rnorm(11, 60, 4)
        sample3 <- rnorm(11, 60, 4)
        m = matrix(c(sample1, sample2, sample3),
            11, 3)
        within <- matrix(rep(NA, 33), 11, 3)
        for (j in 1:3) {
            for (i in 1:11) {
                within[i, j] <- m[i, j] - mean(m[,
                    j])
            }
        }
        ms.withins[n] <- sum(within^2)/(33 - 3)
        between <- matrix(rep(NA, 33), 11, 3)
        for (j in 1:3) {
            for (i in 1:11) {
```

```
          between[i, j] <- mean(m[, j]) - mean(m)
     }
  }
  ms.betweens[n] <- sum(between^2)/(3 - 1)
  fs[n] <- (sum(between^2)/(3 - 1))/(sum(within^2)/(33 -
     3))
}
```

When we display the distribution of MS-within (Figure 5.3):

```
> plot(density(ms.withins), main = "", xlab = "",
    ylab = "")
```

its shape should not be surprising to you. Variance is an unbiased estimator. Here, we are just taking the variance of each of the three samples, adding them up and dividing by $N - 1$: we're pooling variances to get an estimate of the population variance. Each single sample's variance is an unbiased estimator of the population, so it's no surprise that the pooled variance here is also an unbiased estimator.

Now, look at the distribution of MS-between for the three identical populations (Figure 5.4).

```
> plot(density(ms.betweens), main = "", xlab = "",
    ylab = "")
```

When the three populations have the same mean (60) and the same variance (16), the mean of the sampling distribution of MS-within (15.875) points to the population variance and, although the shape of the distribution differs, the mean of the sampling distribution of MS-between (15.907) also points to the population variance.

The key idea of ANOVA is this: when the groups' means are in fact identical, the variance of these two distributions is very close to the population variance (assuming, as we are, that the group variances are identical).

When the null hypothesis is true (group means are identical), these two statistics (MS-between, MS-within) are unbiased estimators of the same number—the population variance. (Keep in mind too that the three populations' *variances* were assumed to be equal in the simulations. We will return to this below.)

5.3.2 The F-distribution

We have seen that when the null hypothesis is in fact true MS-within and MS-between are indeed equal: between-group variance is completely attributable to within-group variance. We also pointed out that an equivalent way to express the null hypothesis is that their ratio equals one. It turns out

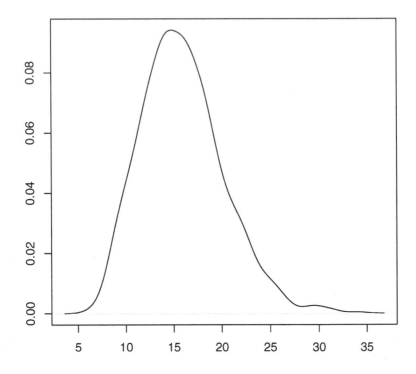

Fig. 5.3 The sampling distribution of MS-within (three identical populations, identical standard deviation).

that if we use this ratio as our statistic, the result is itself a test-statistic. The value we compute for a particular sample using this statistic—the F-STATISTIC—is precisely analogous to a t-statistic; and the accompanying sampling distribution—the F-distribution—can be used precisely like a t-curve to compute the p-value for a result.

We computed these ratios during the above simulation, and now display the sampling distribution (Figure 5.5):

```
> plot(density(fs), xlim = range(0, 8), main = "",
        xlab = "", ylab = "")
```

If the null hypothesis is true, the distribution is centered around 1:

```
> mean(fs)
```

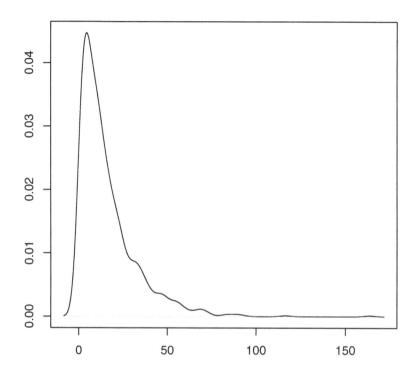

Fig. 5.4 The sampling distribution of MS-between (three identical populations, identical standard deviation).

[1] 1.090634

What is the p-value of this ratio for our particular data-set? We need two things: the relevant F-value, and the precise F-distribution (which our 1000 replicate simulation approximates: i.e. the form this distribution takes as the number of replicates approaches infinity). Recall that the t-distribution is a family of curves defined by one parameter (the degrees of freedom, DF). As we have seen, the ratio that defines the F-statistic involves one DF for the numerator, and another DF for the denominator. The F-distribution is a family of curves defined by these two parameters. We refer to a particular F-distribution as $F(DF_1, DF_2)$ where DF_1 is the degrees of freedom of the numerator (MS-between) and DF_2 is the degrees of freedom of the denominator (MS-within).

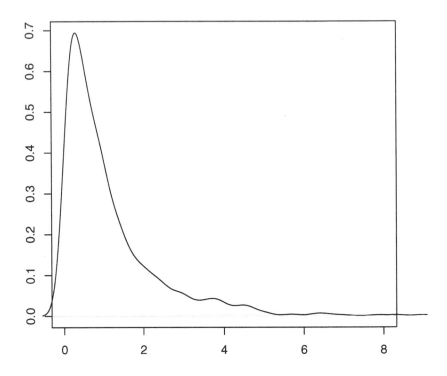

Fig. 5.5 The sampling distribution of the F-ratio when the group means are identical.

In our current example, $DF_1 = 3 - 1$, $DF_2 = 9 - 3$. We can compare the simulated and exact distributions:

```
> multiplot(1, 2)
> title <- "Simulated"
> plot(density(fs), xlim = range(0, 8), main = title,
       xlab = "", ylab = "")
> title <- "Exact"
> plot(function(x) df(x, 2, 6), -0.1, 8, main = title,
       xlab = "", ylab = "")
```

The result is shown in Figure 5.6.

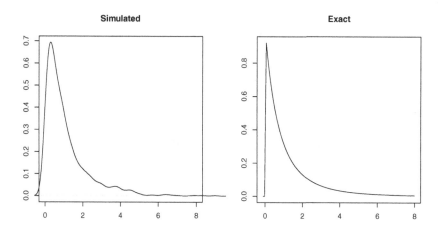

Fig. 5.6 The sampling distribution of $F(2,6)$ both simulated and exact.

We now need the F-statistic for our sample:

```
> ms.between

[1] 12

> ms.within

[1] 14

> ms.between/ms.within

[1] 0.8571429
```

Given our *single sample*'s F-ratio, we can now ask (as in the t-test): What is the probability of getting an F-ratio as far from 1 as this, or further, *assuming that the null hypothesis is true*? This is the p-value of the ANOVA test.

As in the t-test, we integrate the area under the curve corresponding to these extreme values:

```
> integrate(function(x) df(x, 2, 6), 0.8571, Inf)

0.4705232 with absolute error < 1.2e-06
```

The F-statistic for this sample does not provide enough evidence to reject the null hypothesis. The probability of obtaining an F-ratio of 0.8571 or higher by chance when the population means are identical is quite high (0.47).

5.3.3 ANOVA in R

In order to use the built-in ANOVA function in R, we need to wrap the data in a data.frame:

```
> scores <- c(9, 1, 2, 10, 2, 6, 1, 5, 0)
> subj <- paste("s", rep(c(1:9), 1), sep = "")
> group <- paste("g", rep(c(1:3), 1, each = 3),
      sep = "")
> data1 <- data.frame(scores, group, subj)
```

And now, we take a look at the R output for ANOVA:

```
> aov.fm <- aov(scores ~ group + Error(subj), data1)
> summary(aov.fm)
```

```
Error: subj
          Df Sum Sq Mean Sq F value Pr(>F)
group      2     24     12  0.8571 0.4705
Residuals  6     84     14
```

We see exactly the same MS-between. MS-within, F-value and p-value as previously computed.

5.3.4 MS-within, Three Non-identical Populations

Although hypothesis testing is performed using the sampling distribution of the null hypothesis, it is instructive to see what happens to these statistics when we have three populations with different means (still assuming identical population variance). That is, the null hypothesis is now in fact false. Let's plot the distributions of MS-within and MS-between in this case (Figure 5.7).

```
> ms.withins <- rep(NA, 1000)
> ms.betweens <- rep(NA, 1000)
> for (n in 1:1000) {
      sample1 <- rnorm(11, 58, 4)
      sample2 <- rnorm(11, 60, 4)
      sample3 <- rnorm(11, 62, 4)
      m = matrix(c(sample1, sample2, sample3),
          11, 3)
      within <- matrix(rep(NA, 33), 11, 3)
      for (j in 1:3) {
          for (i in 1:11) {
              within[i, j] <- m[i, j] - mean(m[,
                  j])
```

```
            }
        }
        ms.withins[n] <- sum(within^2)/(33 - 3)
        between <- matrix(rep(NA, 33), 11, 3)
        for (j in 1:3) {
            for (i in 1:11) {
                between[i, j] <- mean(m[, j]) - mean(m)
            }
        }
        ms.betweens[n] <- sum(between^2)/(3 - 1)
    }
> mean(ms.withins)

[1] 16.13331

> multiplot(1, 2)
> plot(density(ms.withins), main = "MS-within",
        xlab = "")
> plot(density(ms.betweens), main = "MS-between",
        xlab = "")
```

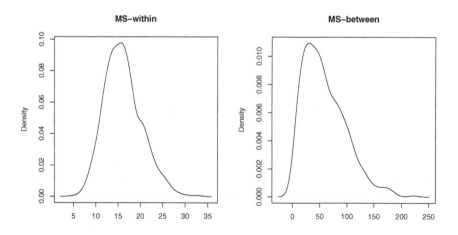

Fig. 5.7 The sampling distribution of MS-within (left) and MS-between (right), three different group means and identical variances.

The mean of the MS-withins is more or less the same as when the three populations' means were identical. Why didn't the mean of MS-within change much in this case, where the population means differ? Reflect upon MS-within for a minute:

$$MS_{within} = \frac{\sum\limits_{j=1}^{I} \sum\limits_{i=1}^{n_j} (x_{ij} - \bar{x}_j)^2}{N - I} \tag{5.53}$$

MS-within is computing the spread about the mean in each sample: the location of the mean in that sample is irrelevant. As long as the population variances remain identical, MS-within will always estimate this variance in an unbiased manner.

So, MS-within is an essentially invariant reference number for a comparison of populations with the same variances. Now let's look at how MS-between behaves with non-identical means in the three populations.

To summarize, the mean of the sampling distribution of MS-within is an invariant reference number for a comparison of populations with the same variances. The mean of the sampling distribution of MS-between is nearly identical to that of MS-within if the null hypothesis is true (identical population means). And this is the F-ratio, $\frac{MS_{between}}{MS_{within}} \approx 1$. If the null hypothesis is in fact false (if the population means differ), then it's highly likely that MS-between is greater than MS-within, and that the F-ratio is significantly greater than 1.

When population means actually differ, for a *given* sample it is possible that MS-between is lower (close to the populations' variance) and that MS-within is higher than the populations' variances. But, because of the shapes of sampling distributions we just saw, the probability of each of these events happening is low, and therefore the co-occurrence of both these events is even less probable.

5.3.5 The F-distribution with Unequal Variances

Note that in all of the above simulations, the variances of the individual groups were identical as assumed by the ANOVA model. What happens when the null hypothesis is true (the population means are identical) but the population variances are not identical? As the variances diverge, the assumptions of the model are compromised and different inference methods are required. A common rule of thumb is that the results of ANOVA will be approximately correct if *the largest standard deviation is less than twice the smallest standard deviation.*

You should now be in a position to evaluate such a claim, by simulating the case where the population means are identical, but the variances differ.

In the next section, we discuss ANOVA at a more advanced level, and go into some detail about how R is used for ANOVA calculations. We also relate ANOVA to the linear model in more detail in chapter 6.

5.4 ANOVA as a Linear Model

You might at this point ask: can we do the ANOVA based on our original null and alternative hypotheses? Let's review what these were:

$$H_0 : x_{ij} = \mu + \varepsilon_{ij} \tag{5.54}$$

$$H_a : x_{ij} = \mu + \alpha_j + \varepsilon_{ij} \tag{5.55}$$

In fact, the MS-between and MS-within method we used is identical to the ANOVA based on the linear models above.

Take a look at this output:

```
> (aov.fm)

Call:
aov(formula = scores ~ group + Error(subj), data = data1)

Grand Mean: 4

Stratum 1: subj

Terms:
                group Residuals
Sum of Squares     24        84
Deg. of Freedom     2         6

Residual standard error: 3.741657
Estimated effects may be unbalanced
```

Towards the beginning, the ANOVA output tells us what the formula for the calculations is:

```
Call:
aov(formula = scores ~ group + Error(subj), data = data1)
```

This is almost literally the alternative hypothesis as a system of linear equations. The only thing missing in the formula above is the term for the grand mean:

$$H_a : x_{ij} = \mu + \alpha_j + \varepsilon_{ij} \tag{5.56}$$

But the alternative hypothesis above is what R is actually using for computation. To see this, let's unpack ANOVA in R further. In reality, the ANOVA call in R is actually doing computations based on a bunch of linear equations, one for each subject. Let's see if we can squeeze this information out of R.

First we are going to fit a linear model (with the function `lm`), and then examine the underlying equations. The code you see below seems obscure at first, but all will become clear in a moment.

```
> lm.fm <- lm(scores ~ group, data = data1)
> (mm.fm <- model.matrix(lm.fm))[1:8, ]

  (Intercept) groupg2 groupg3
1           1       0       0
2           1       0       0
3           1       0       0
4           1       1       0
5           1       1       0
6           1       1       0
7           1       0       1
8           1       0       1

> (cf.fm <- coefficients(lm.fm))

(Intercept)      groupg2      groupg3
          4            2           -2

> (res.fm <- residuals(lm.fm))

 1  2  3  4  5  6  7  8  9
 5 -3 -2  4 -4  0 -1  3 -2

> (cbind(mm.fm, res.fm))

  (Intercept) groupg2 groupg3 res.fm
1           1       0       0      5
2           1       0       0     -3
3           1       0       0     -2
4           1       1       0      4
5           1       1       0     -4
6           1       1       0      0
7           1       0       1     -1
8           1       0       1      3
9           1       0       1     -2
```

The matrix generated by the last call above shows almost all of the terms of the nine equations corresponding to each subject. See if you can discover the relationship between cf.fm, mm.fm, and res.fm. To help you along, we provide the answer below:

$$9 = 4 \times 1 + 2 \times 0 + -2 \times 0 + 5.0e + 00 \tag{5.57}$$
$$1 = 4 \times 1 + 2 \times 0 + -2 \times 0 + -3.0e + 00 \tag{5.58}$$
$$2 = 4 \times 1 + 2 \times 0 + -2 \times 0 + -2.0e + 00 \tag{5.59}$$
$$10 = 4 \times 1 + 2 \times 1 + -2 \times 0 + 4.0e + 00 \tag{5.60}$$
$$2 = 4 \times 1 + 2 \times 1 + -2 \times 0 + -4.0e + 00 \tag{5.61}$$
$$6 = 4 \times 1 + 2 \times 1 + -2 \times 0 + 1.5e - 16 \tag{5.62}$$
$$1 = 4 \times 1 + 2 \times 0 + -2 \times 1 + -1.0e + 00 \tag{5.63}$$
$$5 = 4 \times 1 + 2 \times 0 + -2 \times 1 + 3.0e + 00 \tag{5.64}$$
$$0 = 4 \times 1 + 2 \times 0 + -2 \times 1 + -2.0e + 00 \tag{5.65}$$

Each of these equations gives you the observed score of each subject as a function of the grand mean, the effect of each factor, and the error due to the subject in question. These equations are the 'exploded' form of the compact one we saw earlier:

$$x_{ij} = \mu + \alpha_j + \varepsilon_{ij} \tag{5.66}$$

The connection between the system of equations above and this compact-form equation is not obvious. To highlight the connection, we can restate the system of equations above as a giant matrix-based equation:

$$Y_i = \beta_0 X_{0_i} + \beta_1 X_{1_i} + \beta_2 X_{2_i} + \varepsilon_i \tag{5.67}$$

- Y_i is the matrix containing the scores of all the 9 subjects.

  ```
  > (data1$scores)

  [1]  9  1  2 10  2  6  1  5  0
  ```

- X_{0_i} is the intercept column of 1s in the mm.fm matrix.

  ```
  > (mm.fm[, 1])

  1 2 3 4 5 6 7 8 9
  1 1 1 1 1 1 1 1 1
  ```

- X_{1_i} and X_{2_i} are dummy variables that help us code each subject as being in group 1, 2, or 3 (see below for an explanation).

  ```
  > (mm.fm[, 2:3])

    groupg2 groupg3
  1       0       0
  2       0       0
  3       0       0
  4       1       0
  5       1       0
  ```

```
6        1        0
7        0        1
8        0        1
9        0        1
```

- β_0 is the grand mean (the first element of cf.fm).

  ```
  > (cf.fm[1])
  ```

  ```
  (Intercept)
            4
  ```

- β_1 is the effect of group 2 (the second element of cf.fm).

  ```
  > (cf.fm[2])
  ```

  ```
  groupg2
        2
  ```

- β_2 is the effect of group 3 (the third element of cf.fm).

  ```
  > (cf.fm[3])
  ```

  ```
  groupg3
       -2
  ```

- Exercise: What is ε_i?

Note how the three α components (each corresponding to one of the three groups) are expressed in the system of linear equations above. To figure this out, look at the model matrix output from the linear model once again:

```
> (mm.fm)[, 2:3]
```

```
  groupg2 groupg3
1       0       0
2       0       0
3       0       0
4       1       0
5       1       0
6       1       0
7       0       1
8       0       1
9       0       1
```

Notice that the second and third columns uniquely classify each of the 9 rows as corresponding to subjects 1-9. Subject 1 has group g2=0, and group g3=0, same for subjects 2 and 3: i.e. these subjects are neither in group 2 or 3, they are in group 1; and so on. This kind of coding is called DUMMY CODING.

You can ask R to compute an ANOVA using this linear model. Compare the output of the anova function (which uses the `lm` output) with the earlier anova we had found using the `aov` function:

```
> anova(lm.fm)

Analysis of Variance Table

Response: scores
          Df Sum Sq Mean Sq F value Pr(>F)
group      2     24      12  0.8571 0.4705
Residuals  6     84      14

> summary(aov.fm)

Error: subj
          Df Sum Sq Mean Sq F value Pr(>F)
group      2     24      12  0.8571 0.4705
Residuals  6     84      14
```

Problems

5.1. [Do all computations first explicitly and then compare your result with computations using R's t.test function.]

Given three second-language vocabulary learning methods (I, II, III), three subjects are assigned to each method. The relative effectiveness of learning methods is evaluated on some scale by scoring the increase in vocabulary after using the method.

Table 5.10 Fictional dataset from Rietveld and van Hout (2005).

	Group I	Group II	Group III
	9	10	1
	1	2	5
	2	6	0
$\bar{x}$	4	6	2

Evaluate the research question: Is there any difference in the three learning methods? Do three pairwise t-tests. Can one conclude anything from the results? If so, what? If nothing, why not?

Now do the three t-tests to evaluate the same research question with Table 5.11 from the same setup:

Table 5.11 Fictional dataset from Rietveld and van Hout 2005.

	Group I	Group II	Group III
	3	7	1
	4	6	2
	5	5	3
$\bar{x}$	4	6	2

Is any comparison significant? Can we conclude anything this time regarding the research question? If any of the tests are significant, is the p-value low enough that we can reject that particular null hypothesis at $\alpha = .05$?

Note: you will need either to look up a t-test table from a statistics textbook, or you can use R to ask: what's the critical t for n degrees of freedom at an alpha level of 0.05? The R command for answering this question is `qt(.975,n)`.

5.2. Prove that: $\sum\limits_{j=1}^{I}\sum\limits_{i=1}^{n_j}(x_{ij}-\bar{x})^2 = \sum\limits_{j=1}^{I}\sum\limits_{i=1}^{n_j}(\bar{x}_j-\bar{x})^2 + \sum\limits_{j=1}^{I}\sum\limits_{i=1}^{n_j}(x_{ij}-\bar{x}_j)^2$

5.3. Compute the mean squares and F-value for the second dataset in the same manner (explicitly, using the `anova` command). Isolate the underlying linear equations as we did above and identify all of the coefficients and terms in the general form of the equation:

$$Y_i = \beta_0 X_{0_i} + \beta_1 X_{1_i} + \beta_2 X_{2_i} + \varepsilon_i \qquad (5.68)$$

Here's the data to get you started:

```
> scores <- c(3, 4, 5, 7, 6, 5, 1, 2, 3)
> subj <- paste("s", rep(c(1:9), 1), sep = "")
> group <- paste("g", rep(c(1:3), 1, each = 3),
        sep = "")
> data2 <- data.frame(scores, group, subj)
```

Chapter 6
Bivariate Statistics and Linear Models

So far we've been studying UNIVARIATE STATISTICS; for example, for each individual in a population, we take a single measurement, height, age, etc. We combine these into a sample and compute a statistic: mean, variance, or some function of the variance. Now we consider the scenario where, *for each individual* in a population, we have two values: age *and* height, midterm *and* final exam result, etc. In such a situation we can, of course, treat each dimension independently, and compute the same univariate statistics as before. But the reason we measure two values is to assess the CORRELATION between them, and for this, we require 'two-dimensional' or BIVARIATE STATISTICS. As a concrete example, here are some (allegedly real) homework, midterm, and final exam scores from a statistics course (Faraway's dataset `stat500`).

```
> library(faraway)
> data(stat500)
> head(stat500, n = 2)

  midterm final   hw total
1    24.5  26.0 28.5  79.0
2    22.5  24.5 28.2  75.2
```

The research question is: can we predict the performance of students on the final exam from their midterm scores? Is there any correlation between them? Consider first a trivial variant of such a question: can we predict the performance of students on the final exam from their *final exam* scores? Although this example is totally trivial, it does allow us to visualize what a perfect correlation looks like, and makes the point that it is *perfectly linear*.

First, we create local copies of the `final` and `midterm` scores, since we will need to access these vectors quite often:

```
> final <- stat500$final
> midterm <- stat500$midterm
```

Then we plot the final scores (Figure 6.1).

S. Vasishht, M. Broe, *The Foundations of Statistics: A Simulation-based Approach*, DOI 10.1007/978-3-642-16313-5_6,
© Springer-Verlag Berlin Heidelberg 2011

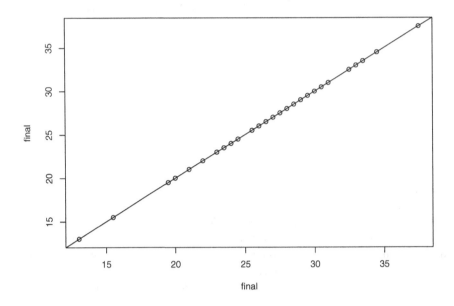

Fig. 6.1 Final scores perfectly predict final scores.

The research question is however whether midterm scores can predict final-exam scores. Let's plot this relationship, which is not quite so perfect (Figure 6.2):

```
> plot(final ~ midterm, xlim = c(0, 35), ylim = c(0,
        35))
> abline(0, 1)
```

Notice that we can express the (univariate) means of the midterm and final by two lines, which cross at a certain point (Figure 6.3):

```
> x <- mean(midterm)
> sdx <- sd(midterm)
> y <- mean(final)
> sdy <- sd(final)
> plot(final ~ midterm)
> arrows(x, min(final), x, max(final), code = 0)
> arrows(min(midterm), y, max(midterm), y, code = 0)
```

In order to predict the final score from the midterm, we need to somehow summarize the relationship between the midterm and final scores. One approach is to ask: 'how linear' is the relationship? We have seen that a perfect correlation is perfectly linear, so an imperfect correlation will be 'imperfectly

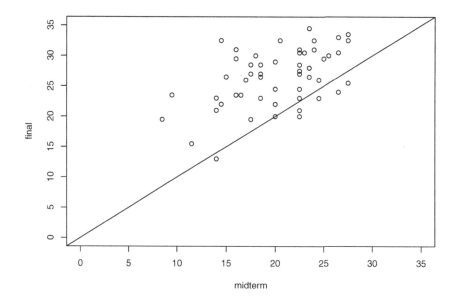

Fig. 6.2 The relationship between midterm and final scores, with the perfectly linear relationship from Figure 6.1 superimposed on the data.

linear.' Be careful, however. We are *not* asking if there is some non-linear shape that expresses the correlation better than a straight line. Rather we are asking, when we *do* fit the best straight line we can to the data, how good is that fit? We will *always* be able to fit a straight line; in fact, we will always be able to fit an unambiguously *best* straight line. Once we do that, how perfectly/imperfectly does that line summarize the data? We are concerned here only with SIMPLE LINEAR, as opposed to NONLINEAR, models. We will see that this line of best fit is readily interpreted as the two-dimensional mean of the data.

When working with two dimensions, which (potentially) use very different units and have very different ranges, it often aids visualization and simplifies the math to agree on some common scale for both. This also facilitates comparisons between different studies. We begin by standardizing the final and midterm scores (i.e., converting the scores to z-scores so that they have mean 0 and standard deviation 1). See Figure 6.4.

```
> scaledstat500 <- data.frame(scale(stat500))
```

Then, we make back-up copies of the final and midterm scores, using the standardized scores instead:

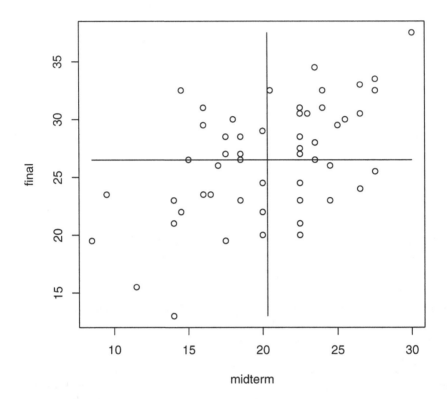

Fig. 6.3 The means of the midterm and final scores.

```
> final.original <- final
> midterm.original <- midterm
> final <- scaledstat500$final
> midterm <- scaledstat500$midterm
```

Finally, we plot the standardized final scores against the standardized midterm scores (Figure 6.4):

```
> plot(final ~ midterm)
> arrows(mean(midterm), min(final), mean(midterm),
      max(final), code = 0)
> arrows(min(midterm), mean(final), max(midterm),
      mean(final), code = 0)
> text(1, 2, labels = expression(x[i] %*% y[i]),
      cex = 1.2)
```

```
> text(1.5, 2, labels = c(" = +ve"), cex = 1.2)
> text(1, -2, labels = expression(x[i] %*% -y[i]),
      cex = 1.2)
> text(1.55, -2, labels = c(" = -ve"), cex = 1.2)
> text(-1.1, -2, labels = expression(-x[i] %*%
      -y[i]), cex = 1.2)
> text(-0.5, -2, labels = c(" = +ve"), cex = 1.2)
> text(-1, 2, labels = expression(-x[i] %*% y[i]),
      cex = 1.2)
> text(-0.5, 2, labels = c(" = -ve"), cex = 1.2)
```

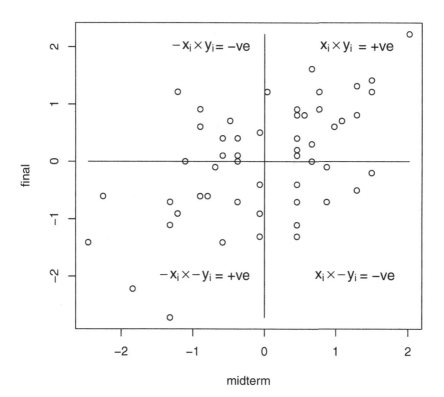

Fig. 6.4 The standardized scores of the midterm and final scores.

If we now multiply each pair of standardized values (x, y), the product will differ in sign depending on which quadrant of this standardized space the pair occupies. Further, the sum of all of these products will be positive just in case more points lie in the first and third (positive) quadrants than in the other two (and there are no major extreme values in the second and fourth quadrants).

Correlation r is defined as:

$$r = \frac{\sum\limits_{i=1}^{n} (z_{x_i} \times z_{y_i})}{n-1} \tag{6.1}$$

where z_{x_i} refers to the z-score of x_i, etc.

Let's do a quick sanity check here. We compute the correlation (as defined above) explicitly, and then use R's built-in function cor() to verify our computation.

```
> sum(final * midterm)/(length(final) - 1)

[1] 0.5452277

> cor(midterm, final)

[1] 0.5452277
```

The positive value correlation is telling us that the majority of the (x, y) pairs are located in the first and third quadrants. So we can say, roughly, that the higher the midterm score, the higher the final score is likely to be. We will formulate a more precise expression of the relation below.

Here's a more subtle question: Given that some particular mid-term score is above average by a certain amount, is the corresponding final score above average by that same amount? More? Less? This issue caught the attention of Francis Galton when studying the resemblance between parents and offspring. If the parent is taller than average to some extent, the child is expected to be taller, but is the child taller *to the same extent*? More? Less? (Galton's parent-child data is available from the R library UsingR.)

```
> library(UsingR)
> data(galton)
> parent <- galton$parent
> child <- galton$child
> gx <- mean(parent)
> gsdx <- sd(parent)
> gy <- mean(child)
> gsdy <- sd(child)
> plot(child ~ parent)
> arrows(gx, min(child), gx, max(child), code = 0)
> arrows(min(parent), gy, max(parent), gy, code = 0)
```

Theoretically, populations could diverge from, stay constant with respect to, or approach the mean in the next generation. Galton found that they

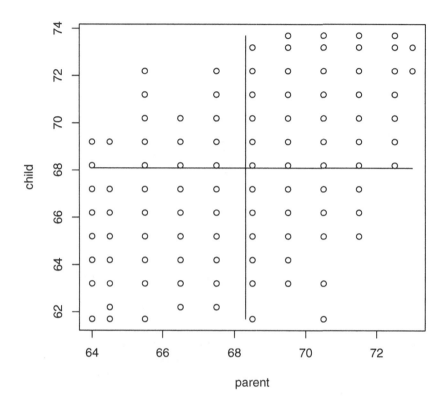

Fig. 6.5 Galton's parent-child data.

approach the mean (if they diverged, the tall would get taller, and taller, and taller...). In Galton's words the values *regress* towards the mean. It is for this reason that the line of best fit is termed the REGRESSION LINE.

Let's return to the midterm-final example to see how we can establish what happens in that case. We'll find that here too the final score regresses towards the mean. Figure 6.6 shows the plot again.

How do we get a handle on this relationship? It's quite hard to see by merely inspecting the plot. Let's focus on a subset of the data in a vertical strip that contains midterm scores that are about 1 SD above the mean, and a subset that contains scores about 1 SD below the mean, and then examine the mean of the final scores for those subsets:

```
> (one.SD.above <- subset(scaledstat500, 0.9 <
      midterm & midterm < 1.1))
```

```
     midterm      final        hw      total
9  0.9763516 0.6074129  0.2640458 0.8604525
46 1.0806222 0.7083426 -1.7356125 0.1718759
```

```
> mean(one.SD.above$final)
```

```
[1] 0.6578777
```

```
> (one.SD.below <- subset(scaledstat500, -0.9 >
      midterm & midterm > -1.1))
```

```
      midterm      final         hw      total
6  -0.9005185  0.9102018  0.31403725  0.1423654
39 -0.9005185  0.6074129 -3.73527088 -1.5987496
41 -0.9005185 -0.6037428 -0.01090723 -0.7232737
```

```
> mean(one.SD.below$final)
```

```
[1] 0.304624
```

So, when the midterm score is 1 SD above the midterm mean, the final score is only 0.65 SD above the mean of the final score. When the midterm score is 1 SD below the midterm mean, the final score is only 0.30 SD below the final mean. The final scores are regressing towards the mean. They would not be regressing towards the mean if the ratio of final to midterm were 1:1, for example.

To make this point more concrete, let's split the difference between these two results and draw the line $y = 0.475 \times x$, and compare it with the line where a 1 SD change in midterm score results in a 1 SD change in the final score (in the same direction): $y = x$, Figure 6.6. Here you see visually what is meant by regression to the mean.

```
> plot(final ~ midterm)
> arrows(mean(midterm), min(final), mean(midterm),
      max(final), code = 0)
> arrows(min(midterm), mean(final), max(midterm),
      mean(final), code = 0)
> abline(0, 1, col = "black")
> abline(0, 0.5452, col = "black", lty = 2)
> text(1.5, 2, labels = c("1:1 ratio of change"),
      col = "black")
> text(1.45, 0.3, labels = c("0.5:1 ratio of change"),
      col = "black")
```

When we average over a strip, we find a mean—the center of gravity, as it were—for those points in the strip. We did this twice, and we can imagine doing this across the entire range of midterm values, finding the means of final values, and then joining these means up to make a line which is intuitively the 'center' of all of the data points. Recall now the insight

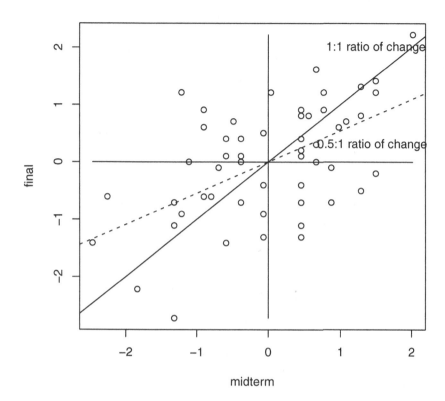

Fig. 6.6 Comparing the final and midterm data's rate of change of 0.5:1 with a situation where the rate of change is 1:1.

we developed in Chapter 1: the mean minimizes variance of the errors from the mean. The method of LEAST SQUARES is a mathematical technique that identifies an unknown mean by solving a set of equations in such a way as to minimize this variance. The technique applies in two dimensions as well, and is a more sophisticated and mathematically exact way of achieving the same thing what we imagined doing piecemeal above: finding the line of best fit, the two dimensional mean. The R command for computing this line of best fit is lm:

```
> summary(lm(final ~ midterm))

Call:
lm(formula = final ~ midterm)
```

```
Residuals:
    Min      1Q  Median      3Q     Max
-2.0049 -0.5363  0.1064  0.6024  1.8745


Coefficients:
              Estimate Std. Error  t value Pr(>|t|)
(Intercept) -3.647e-16  1.141e-01 -3.20e-15        1
midterm      5.452e-01  1.151e-01    4.735 1.67e-05


Residual standard error: 0.8462 on 53 degrees of freedom
Multiple R-squared: 0.2973,      Adjusted R-squared: 0.284
F-statistic: 22.42 on 1 and 53 DF,  p-value: 1.675e-05
```

The above command shows us that this least squares line is as in equation 6 (recall that $5.452e-01$ is the standard exponential notation that translates to 0.542). The intercept is effectively zero, as we would expect in standardized data, and the slope is the rate of change.

$$\text{final} = 0.5452 \times \text{midterm} \tag{6.2}$$

Recall that we'd previously calculated that $r = 0.5452$:

$$\frac{z_y}{z_x} = 0.5452 \tag{6.3}$$

The correlation is thus the slope of the least-squares regression line when we measure the variables in standardized units. To get the slope in the unstandardized data, we can fit a least-squares line on the original data (Figure 6.7).

```
> final <- final.original
> midterm <- midterm.original
> lm.stat500 <- lm(final ~ midterm)
> plot(final ~ midterm)
> abline(lm.stat500)
> text(15, 24, labels = c("y = 15.0462 + 0.5633x"),
       cex = 1.2)
```

Now, the intercept and slope are different:

$$\hat{y} = 15.0462 + 0.5633 \times x \tag{6.4}$$

Notice the important asymmetry between the x and y dimensions in regression analysis. When we fit the line to the data, we minimize distances from the line *in the vertical dimension*, not in the horizontal dimension, and not orthogonal to the line itself. In our example, midterm is termed the *explanatory* variable, final is called the *response* variable. We express the response variable as a function of the explanatory variable; we can use the regression

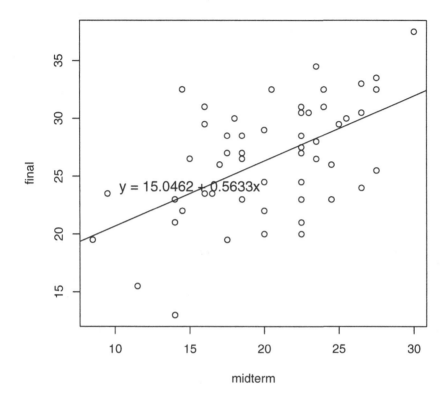

Fig. 6.7 The regression line in the unstandardized data.

line to *predict* the value of the response y for a given value of x. In our example we computed the *regression of final on midterm*. The regression of *midterm on final* would give a different slope. Because we use regression to predict the value of y, it is errors in the vertical direction that are of interest. The intercept is the predicted final score when the midterm score is 0, and the slope is the increase in the final score as the midterm score increases by 1.

6.1 Variance and Hypothesis Testing for Regression

If the regression line is the 'bivariate mean,' what is the corresponding measure of variance in two dimensions?

Look at this equation again:

$$\hat{y} = 15.0462 + 0.5633 \times x \qquad (6.5)$$

For any x, $\hat{y}$ is the predicted value: in reality there is considerable deviation from this $\hat{y}$. Call $y_i - \hat{y}_i$ the RESIDUAL ERROR. If the predicted value was perfect there would be no residual error. If we square and sum these errors, we have a measure strictly analogous to variance computed from a single mean.

What we are doing then is breaking apart any actual observed value y_i into the model parameters (β_0 and β_1) plus an error (or residual) term ε_i:

$$y_i = \beta_0 + \beta_1 x + \varepsilon_i \qquad (6.6)$$
$$= \hat{y}_i + \varepsilon_i \qquad (6.7)$$

There is a very strong analogy here to the way we interpreted variance in ANOVA. To see this, imagine creating discrete strips all across the `midterm` variable, and imagine each of these as a sample from a different population. For each we find the group mean. In ANOVA, to compute MS-within, we found the difference between each observed value and the local group mean—all the error terms or residuals—squared and summed them to get SS-within and divided by the appropriate degrees of freedom (DF) to find the mean square. In regression, we minimized the sum of squared residuals when we fit the regression line—that's what the method of least squares does! These residuals are computed by the function `lm()` behind the scenes, but can be accessed explicitly:

```
> residuals <- residuals(lm(final ~ midterm))
> tail(residuals, n = 5)
```

```
        51          52          53          54          55
-0.7198741   3.0046930  -4.7198741   3.0270235  -0.2831497
```

If we square them and sum them and divide by the appropriate DF, we have the regression equivalent of MS-within. Just as in ANOVA, the total DF is $n - 1$, here $55 - 1$. The regression equivalent of 'between group' variance is the systematic variation modeled by the slope of the regression line. This is a function of the value of the single explanatory variable x, and has DF $2 - 1 = 1$. Finally, since the total DF is the sum of the other two, we have DF of the residuals $n - 2$ (i.e., 53). We can now compute the residual SS and MS:

```
> sum(residuals^2)
```

```
[1] 931.2854
```

```
> ms.residuals <- sum(residuals^2)/53
> ms.residuals
```

```
[1] 17.57142
```

For every residual we have a value on the regression line which corresponds to the fitted mean value. We can access these as well:

```
> fitted.values <- fitted.values(lm(final ~ midterm))
> tail(fitted.values)
```

50	51	52	53	54	55
25.18513	27.71987	23.49531	27.71987	29.97298	28.28315

If we subtract these from the grand mean, and square and sum them, we have the regression equivalents of SS-between and MS-between (which are equal, since the DF is 1).

```
> mean(final)
```

```
[1] 26.49091
```

```
> ms.fitted <- sum((fitted.values - mean(final))^2)/1
```

In summary, if we let $\bar{y}$ be the grand mean, and $\hat{y}_i$ be the fitted regression mean, we have the following analogies to ANOVA:

ANOVA	Regression
$SS_{\text{total}} = \sum\limits_{j=1}^{I} \sum\limits_{i=1}^{n_j} (x_{ij} - \bar{x})^2$	$SS_{\text{total}} = \Sigma(y_i - \bar{y})^2$
$SS_{\text{between}} = \sum\limits_{j=1}^{I} \sum\limits_{i=1}^{n_j} (\bar{x}_j - \bar{x})^2$	$SS_{\text{fitted}} = \Sigma(\hat{y}_i - \bar{y})^2$
$SS_{\text{within}} = \sum\limits_{j=1}^{I} \sum\limits_{i=1}^{n_j} (x_{ij} - \bar{x}_j)^2$	$SS_{\text{residual}} = \Sigma(y_i - \hat{y}_i)^2$

The null hypothesis for regression is that there is no association or correlation at all between the explanatory variable and the response variable. Thus the x and y values are not linearly related, and we have:

$$H_0 : \beta_1 = 0 \tag{6.8}$$

This is interpreted, just as in ANOVA, as the claim that all of the variance in the data is simply residual variance.

This is the situation we might expect if the x and y values were paired randomly, for example. The degree to which this is *not* the case is the degree to which we have evidence for the alternative hypothesis that there truly is a linear relationship between the variables. Thus, just as in ANOVA, we can compute an F-statistic (the ratio of fitted to residual values), and a p-value using the F distribution with the appropriate DFs.

We can simulate this null hypothesis quite easily in R, just as we did for ANOVA (the function **sample** performs a random permutation of the response values):

```
> f.ratios <- rep(NA, 1000)
> for (i in 1:1000) {
      final.rand <- sample(final)
      lm <- lm(final.rand ~ midterm)
      residuals <- residuals(lm)
      ms.residuals <- sum(residuals^2)/53
      fitted.values <- fitted.values(lm)
      ms.fitted <- sum((fitted.values - mean(final))^2)/1
      f.ratios[i] <- ms.fitted/ms.residuals
  }
> multiplot(1, 2)
> main.title <- "Distribution of \n simulated F-ratios"
> plot(density(f.ratios), xlim = range(0, 8), main = main.title,
      xlab = "", ylab = "")
> main.title <- "Distribution of \n theoretical F-ratios"
> plot(function(x) df(x, 1, 53), -0.1, 8, main = main.title,
      xlab = "", ylab = "")
```

The plots resulting from this code are shown in Figure 6.8.

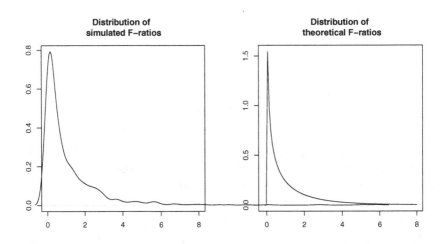

Fig. 6.8 Simulated and theoretical F distributions.

R does precisely this calculation using the output of the linear model (lm) function:

```
> summary(lm.stat500)
```

```
Call:
lm(formula = final ~ midterm)
```

```
Residuals:
   Min     1Q Median    3Q     Max
-9.932 -2.657  0.527  2.984  9.286
```

```
Coefficients:
             Estimate Std. Error t value Pr(>|t|)
(Intercept)  15.0462     2.4822   6.062 1.44e-07
midterm       0.5633     0.1190   4.735 1.67e-05
```

```
Residual standard error: 4.192 on 53 degrees of freedom
Multiple R-squared: 0.2973,        Adjusted R-squared: 0.284
F-statistic: 22.42 on 1 and 53 DF,  p-value: 1.675e-05
```

```
> anova(lm.stat500)
```

```
Analysis of Variance Table
```

```
Response: final
          Df Sum Sq Mean Sq F value    Pr(>F)
midterm    1 393.96  393.96  22.421 1.675e-05
Residuals 53 931.29   17.57
```

As an aside, notice that the above can be done directly in R without fitting the linear model:

```
> summary(aov(final ~ midterm, stat500))
```

```
          Df Sum Sq Mean Sq F value    Pr(>F)
midterm    1 393.96  393.96  22.421 1.675e-05
Residuals 53 931.29   17.57
```

One intriguing thing to notice is that the t-value generated by the `lm` function, 4.735, and the F-score, 22.421 seem to be in a lawful relationship: $t^2 = F$:

```
> 4.735^2
```

```
[1] 22.42023
```

This relationship is discussed in standard textbooks, such as Maxwell & Delaney, 2000, but demonstrating the reason for this identity lies outside the scope of this book. (In additional material provided on the book homepage, we provide a summary of the explanation.)

6.1.1 Sum of Squares and Correlation

Notice that if all observed values y had been on the line, then $y_i = \hat{y}_i$. That is, $SS_{total} = SS_{fitted}$. This was the case in the perfectly linear relationship in Figure 6.1. In this case the regression line accounts for all of the variation.

By contrast, if the observed values of y_i spread out around the predicted values $\hat{y}_i$, then the regression line predicts only part of the variation. We can state this more succinctly:

$$SS_{total} = SS_{fit} + SS_{residual} \tag{6.9}$$

$$\frac{SS_{total}}{SS_{total}} = \frac{SS_{fit}}{SS_{total}} + \frac{SS_{residual}}{SS_{total}} \tag{6.10}$$

$$1 = \frac{SS_{fit}}{SS_{total}} + \frac{SS_{residual}}{SS_{total}} \tag{6.11}$$

Clearly, $\frac{SS_{fit}}{SS_{total}}$ tells us what proportion of the variance the regression equation can predict. This ratio is in fact the square r^2 of the correlation, which we computed earlier. The closer r^2 is to 1, the better the model fit, and the more the variation explained by the data.

Problems

6.1. Determine whether in the Galton dataset the height of the parent predicts the height of their child. Write out the regression equation, and explain what the intercept and slope mean.

6.2. Load the dataset MathAchieve, after downloading it from the book's homepage:

```
> MathAchieve <- read.table("mathachieve.txt")
> colnames(MathAchieve) <- c("School", "Minority",
      "Sex", "SES", "MathAch", "MEANSES")
> head(MathAchieve, n = 3)
```

	School	Minority	Sex	SES	MathAch	MEANSES
1	1224	No	Female	−1.528	5.876	−0.428
2	1224	No	Female	−0.588	19.708	−0.428
3	1224	No	Male	−0.528	20.349	−0.428

This dataset shows math achievement scores and socio-economic status of the students. The details of this study are discussed in Raudenbush & Bryk, 2002. The SES is sometimes negative because it is 'centered': the actual

SES minus the mean SES of a particular school. In other words, an SES of 0 represents the mean SES. Determine whether socio-economic status can predict math achievement. Explain what the intercept and slope mean.

6.3. Load the dataset `beauty.txt` after downloading it from the book's homepage:

```
> beauty <- read.table("beauty.txt", header = TRUE)
> head(beauty)

      beauty evaluation
1  0.2015666        4.3
2 -0.8260813        4.5
3 -0.6603327        3.7
4 -0.7663125        4.3
5  1.4214450        4.4
6  0.5002196        4.2
```

This dataset is discussed in detail in Gelman & Hill, 2007. The dataset has teaching evaluation scores of professors from various universities along with an evaluation of how good they look visually (their 'beauty' level). The research question is whether the beauty-level of a professor is a good predictor of their teaching evaluation scores. Do better-looking professors get better evaluations?

Determine whether beauty level can predict teaching evaluation scores. Given the result that you get, do you think there might be a causal relationship between beauty and teaching evaluations? In other words, given the results of this study, would it be a good idea for teachers to try to look nicer in order to get better teaching evaluations?

Chapter 7
An Introduction to Linear Mixed Models

This chapter introduces linear mixed models at an elementary level. The prerequisite for understanding this presentation is a basic knowledge of the foundational ideas of linear regression discussed in the preceding chapter.

7.1 Introduction

The standard linear model has only one random component, that is, the error or residual term ε_i, and its associated variance $\mathrm{Var}(\varepsilon_i)$. In this chapter we consider a more sophisticated model.

We begin with a dataset discussed in Raudenbush & Bryk, 2002. This dataset, available in the R library **nlme**, contains math achievement scores for students in 160 schools, and also provides the sex, socio-economic status (SES), and minority status of each student.[1]

```
> MathAchieve <- read.table("mathachieve.txt")
> colnames(MathAchieve) <- c("School", "Minority",
       "Sex", "SES", "MathAch", "MEANSES")
> head(MathAchieve, n = 3)

  School Minority    Sex    SES MathAch MEANSES
1   1224       No Female -1.528   5.876  -0.428
2   1224       No Female -0.588  19.708  -0.428
3   1224       No   Male -0.528  20.349  -0.428
```

---

[1] Note: It is not a good idea to load the **nlme** library and then load the **lme4** library in the same session, since the latter supersedes the former; loading both libraries in the same session will result in some functions in **lme4** not functioning as intended since they will be masked by the same (older) function in **nlme**. In order to save the reader the trouble of first loading **nlme** and downloading the data to a text file before proceeding with this chapter, we provide the dataset in the library **vb** accompanying this book, as well as on the book's homepage, http://www.purl.oclc.org/NET/vasishth/VB/.

S. Vasishth, M. Broe, *The Foundations of Statistics: A Simulation-based Approach*, DOI 10.1007/978-3-642-16313-5_7,
© Springer-Verlag Berlin Heidelberg 2011

The SES is sometimes negative because it is 'centered': the actual SES minus the mean SES of a particular school. The reason for this will become clear in a moment.

7.2 Simple Linear Model

Suppose our research question is: Does socio-economic status (SES) correlate with math achievement? A linear model for this dataset would predict math achievement as a function of SES:

$$Y_i = \beta_o + \beta_1 X_i + \varepsilon_i \tag{7.1}$$

```
> lm0 <- lm(MathAch ~ SES, data = MathAchieve)
> summary(lm0)

Call:
lm(formula = MathAch ~ SES, data = MathAchieve)

Residuals:
     Min       1Q   Median       3Q      Max
-19.4382  -4.7580   0.2334   5.0649  15.9007

Coefficients:
            Estimate Std. Error t value Pr(>|t|)
(Intercept) 12.74740    0.07569  168.42   <2e-16
SES          3.18387    0.09712   32.78   <2e-16

Residual standard error: 6.416 on 7183 degrees of freedom
Multiple R-squared: 0.1301,        Adjusted R-squared:  0.13
F-statistic:  1075 on 1 and 7183 DF,  p-value: < 2.2e-16

> coefficients(lm0)

(Intercept)            SES
   12.74740        3.18387
```

The coefficients tell us that the linear model is:

$$Y_i = 12.74 + 3.18 X_i + \varepsilon_i \tag{7.2}$$

where X_i is the SES of each student, and ε_i is the random error associated with each student. The error is assumed to be independent and identically distributed ('iid' for short), and is assumed to have a normal distribution centered around 0 with standard deviation σ (this is usually written as $\varepsilon_i \sim N(0, \sigma^2)$). The term iid means that each error value in the sequence of errors comes from the same probability distribution (hence identical) and each error value is independent of the other values.

The linear model above helps us answer the question about the relationship between SES and math achievement. The math achievement for any student i is predicted by a constant term 12.74, plus a factor 3.18 that is multiplied with the SES of that student, plus some error associated with that student.

But the above model cannot answer some other, perhaps more interesting questions:

1. Do schools with higher mean math achievement also have stronger associations between SES and achievement (than schools with lower mean achievement scores)?
2. Does SES predict math achievement to the same extent in each school? You can guess that this is probably not true, but how to find this out? If SES is not an important predictor in some schools but is in some others, this is potentially an important issue we should not ignore.
3. Suppose schools can be separated out into different types, say Public versus Catholic. After we control for mean SES, do the two school types differ in terms of mean math achievement and the strength of the SES-math achievement relationship?

Linear mixed models help us answer such questions. Our goal in the coming pages is to fit a more articulated linear model, where we have a separate intercept and slope for each school. Remember that the linear model above is fitting a single intercept and slope for all scores; it does not take individual variation into account at all. It is quite likely that the schools differ from each other; if so, our simple model does not adequately model the data.

To see this variability between schools, we focus on data from the first two schools, and plot the regression line for achievement against SES for each school.

Here is the linear model for school 1224:

```
> MathAchieve.1224 <- subset(MathAchieve, School ==
      1224)
> lm1 <- lm(MathAch ~ SES, data = MathAchieve.1224)
> summary(lm1)

Call:
lm(formula = MathAch ~ SES, data = MathAchieve.1224)

Residuals:
    Min      1Q  Median      3Q     Max
-12.849  -6.377  -1.164   6.528  12.491

Coefficients:
            Estimate Std. Error t value Pr(>|t|)
(Intercept)   10.805      1.337   8.081 2.63e-10
```

SES   2.509   1.765   1.421   0.162

```
Residual standard error: 7.51 on 45 degrees of freedom
Multiple R-squared: 0.04295,        Adjusted R-squared: 0.02168
F-statistic:  2.02 on 1 and 45 DF,  p-value: 0.1622
```

```
> coef(lm1)
```

```
(Intercept)          SES
  10.805132    2.508582
```

And here is the linear model for school 1288:

```
> MathAchieve.1288 <- subset(MathAchieve, School ==
      1288)
> lm2 <- lm(MathAch ~ SES, data = MathAchieve.1288)
> summary(lm2)
```

```
Call:
lm(formula = MathAch ~ SES, data = MathAchieve.1288)
```

```
Residuals:
    Min       1Q  Median       3Q      Max
-15.648   -5.700   1.047    4.420    9.415
```

```
Coefficients:
            Estimate Std. Error t value Pr(>|t|)
(Intercept)   13.115       1.387   9.456 2.17e-09
SES            3.255       2.080   1.565    0.131
```

```
Residual standard error: 6.819 on 23 degrees of freedom
Multiple R-squared: 0.09628,        Adjusted R-squared: 0.05699
F-statistic:  2.45 on 1 and 23 DF,  p-value: 0.1312
```

```
> coef(lm2)
```

```
(Intercept)          SES
  13.114937    3.255449
```

The commands above show the computations for the linear model—achievement as a function of SES—for each of the two schools. Let's visualize these two fits:

```
> multiplot(1, 2)
> plot(MathAch ~ SES, data = subset(MathAchieve,
      School == 1224), main = "School 1224")
> abline(lm1$coefficients)
> plot(MathAch ~ SES, data = subset(MathAchieve,
      School == 1288), main = "School 1228")
> abline(lm2$coefficients)
```

Figure 7.1 shows the resulting plot:

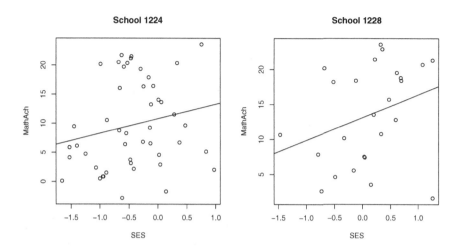

Fig. 7.1 The linear models for schools 1224 and 1288.

A detail to notice: the x-axis is centered around 0. This is because each SES score is centered by subtracting the mean SES score for a school from the raw SES score. The advantage of centering is that it makes the intercept more meaningful (the intercept will now be the mean achievement for the school). Note that one could have centered around the mean SES of students rather than school; that would lead to a different interpretation of the intercept.

So we have fit two regression lines, one for each of the two schools, and for each school the equation looks like this (taking the centering idea into account in the equation):

$$Y_i = \beta_o + \beta_1(X_i - \bar{X}) + \varepsilon_i \tag{7.3}$$

Now, obviously, we can fit separate regression lines for *each* of the schools in the dataset. We can visualize these separate fits quite easily, Figure 7.2.

First we set some sensible defaults for the `lattice` plots we use in this chapter:

```
> scalelist <- list(x = list(alternating = 1),
        y = list(alternating = 1), tck = c(0.5))
```

Then, we define a function for drawing linear regression lines within the plot:

```
> drawfittedline <- function(x, y) {
        panel.xyplot(x, y)
```

```
    panel.lmline(x, y, type = "l", lwd = 1)
}
```

Finally, we use the `xyplot` function from the `lattice` package (Sarkar, 2008) to display the separate regression lines.

```
> library(lattice)
> nine.schools <- unique(MathAchieve$School)[1:9]
> print(xyplot(MathAch ~ SES | factor(School),
        subset(MathAchieve, School %in% nine.schools),
        layout = c(3, 3), cex.axis = 2, xlab = "Student SES",
        ylab = "Math Achievement", panel = drawfittedline,
        scales = scalelist))
```

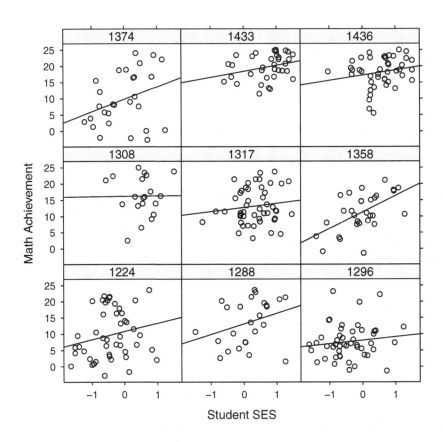

Fig. 7.2 Separate linear models for each school in the dataset (the first nine schools are shown).

The command below shows how to compute separate regression lines in R for each school. If you examine the object lmm1 you will see that it contains the intercepts and slopes for each school. For clarity we show only the two schools' intercepts and slopes computed above. Compare these intercepts and slopes to those we computed earlier: they are identical.

```
> library(lme4)
> lmm1 <- lmList(MathAch ~ SES | School, MathAchieve)
> lmm1$"1224"

Call:
lm(formula = formula, data = data)

Coefficients:
(Intercept)          SES
     10.805         2.509

> lmm1$"1288"

Call:
lm(formula = formula, data = data)

Coefficients:
(Intercept)          SES
     13.115         3.255
```

Notice an important point: we can do a t-test on the list of intercepts and slopes to determine if they are significantly different from zero:

```
> t.test(coef(lmm1)[1])

        One Sample t-test

data:  coef(lmm1)[1]
t = 57.2824, df = 159, p-value < 2.2e-16
alternative hypothesis: true mean is not equal to 0
95 percent confidence interval:
 12.16658 13.03551
sample estimates:
mean of x
 12.60104

> t.test(coef(lmm1)[2])

        One Sample t-test

data:  coef(lmm1)[2]
t = 17.0747, df = 159, p-value < 2.2e-16
alternative hypothesis: true mean is not equal to 0
```

```
95 percent confidence interval:
 1.946981 2.456300
sample estimates:
mean of x
 2.201641
```

The separate regression lines for each school j can be characterized as a single system of equations ($\bar{X}_{.j}$ refers to the mean of school j):

$$Y_{ij} = \beta_{oj} + \beta_{1j}(X_{ij} - \bar{X}_{.j}) + \varepsilon_{ij} \qquad (7.4)$$

We now have a separate intercept and slope for each school: β_{oj}, and β_{1j}. These intercepts and slopes have a variance, and they covary. COVARIANCE is defined as follows:

$$\text{Covariance} = Cov(X,Y) = \frac{\Sigma(X - \bar{x})(Y - \bar{x})}{n - 1}. \qquad (7.5)$$

Covariance allows us to characterize how the values of one variable change as the other variable changes. For example, when one variable increases, the other could also increase (negative covariance); or when one variable increases, the other could decrease (negative covariance); or there could be no such relationship (zero covariance).

Let's name the variances above:

1. $\text{Var}(\beta_{oj}) = \tau_{00}$
2. $\text{Var}(\beta_{1j}) = \tau_{11}$
3. $\text{Cov}(\beta_{0j},\beta_{1j}) = \tau_{01}$

These three τs allow us to compute the correlation between means and slopes (the R code below shows this relationship in R):

$$Cor(\beta_{oj},\beta_{1j}) = \frac{\tau_{01}}{\sqrt{\tau_{00}\tau_{11}}} \qquad (7.6)$$

These correlations are interesting for the following reason. The effectiveness and 'equity' for each school j is described by the pair (β_{0j},β_{1j}). That is, if the intercept for a school has a high value, it is an effective school (in terms of math achievement), and if the slope is small then the school is more equitable across SESs.

We can now ask the following question: is there a relationship between individual school means (i.e., intercepts) and slopes? Are schools that show higher overall effectiveness also more equitable?

Consider the covariance of (β_{0j},β_{1j}). If τ_{01} is positive, this means that increasing effectiveness makes schools less equitable. In R, we can ask this question directly. We can also compute the covariance mentioned above, and verify the relationship between the various τs.

```
> lme1 <- lmList(MathAch ~ SES | School, MathAchieve)
> intercepts <- coef(lme1)[1]
> slopes <- coef(lme1)[2]
> (cov(intercepts, slopes))
```

```
                  SES
(Intercept) 0.3555241
```

```
> (cov(intercepts, slopes)/sqrt(var(intercepts) *
      var(slopes)))
```

```
                   SES
(Intercept) 0.07833762
```

```
> (cor(intercepts, slopes))
```

```
                   SES
(Intercept) 0.07833762
```

It appears that $\tau_{01} = 0.36$. Thus, greater effectiveness in a school correlates with greater inequity across socio-economic status.

Let's also take a graphical look at how the intercepts and slopes across schools relate to each other, Figure 7.3.

```
> intslopes <- data.frame(intercepts, slopes)
> colnames(intslopes) <- c("Intercepts", "Slopes")
> plot(Intercepts ~ Slopes, intslopes)
> lm.intslopes <- lm(Intercepts ~ Slopes, data = intslopes)
> abline(coefficients(lm.intslopes))
```

In this dataset, we also have information about which school is Catholic and which not. We can load a related dataset that provides that information and merge it with the one we have:

```
> MathAchSchool <- read.table("mathachschool.txt")
> colnames(MathAchSchool) <- c("School", "Size",
      "Sector", "PRACAD", "DISCLIM", "HIMINTY",
      "MEANSES")
> MathScores <- merge(MathAchieve, MathAchSchool,
      by = "School")
```

Suppose that we have a hypothesis (two, actually): Catholic schools are more effective and more egalitarian than Public schools (See Raudenbush & Bryk, 2002 for details on why this might be so). How can we find out if these two hypotheses are valid?

Basically, we need to define (i) a model which predicts effectiveness as a function of school type; and (ii) a model which predicts equity as a function of school type. So we need one equation to predict the intercept β_{oj} and another to predict the slope β_{1j}, and we need a way to specify school type. We can

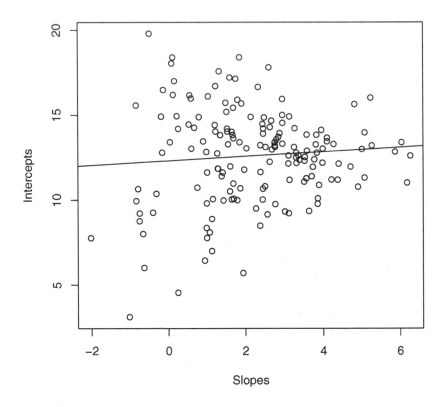

Fig. 7.3 The relationship between intercepts and slopes.

characterize each school j as being Catholic or Public by defining a variable W_j that has value 0 for Catholic school and 1 for Public school.

Note that R has a default dummy coding from the data in such cases:[2]

```
> (contrasts(MathScores$Sector))

          Public
Catholic     0
Public       1
```

The intercept and slope of each school can now be characterized as follows:

$$\beta_{oj} = \gamma_{00} + \gamma_{01} W_j + \varepsilon_{oj} \tag{7.7}$$

---

[2] See the book's homepage (http://www.purl.oclc.org/NET/vasishth/VB/) for more examples of how to code qualitative variables for more complex experimental designs.

and

$$\beta_{1j} = \gamma_{10} + \gamma_{11}W_j + \varepsilon_{1j} \tag{7.8}$$

Here, γ_{00} is the mean achievement for Catholic schools, γ_{01} is the mean achievement difference between Catholic and Public schools; γ_{10} is the average SES achievement slope for Catholic schools; γ_{11} is the mean difference in SES-achievement slopes between Catholic and Public schools; ε_{0j} is the effect of school j on mean achievement holding W_j constant; and ε_{1j} is the effect of school j on the SES-achievement slope holding W_j constant.

Now, we cannot estimate the above two linear models in the usual way; in order to do that the slopes and intercepts would have to have been dependent variables that had been *observed* in the data. The intercepts and slopes we have in the above code are not observed values. So what to do now? Our goal was to use the above equations to evaluate the hypotheses about effectiveness and equity as a function of school type.

Consider the model we saw earlier:

$$Y_{ij} = \beta_{oj} + \beta_{1j}(X_{ij} - \bar{X}_{.j}) + \varepsilon_{ij} \tag{7.9}$$

We can use this equation to assemble one giant predictor equation which shows achievement scores as a function of school type. We can do this by just substituting the equations for the intercepts and slopes.

$$
\begin{aligned}
Y_{ij} =& \beta_{oj} + \beta_{1j}(X_{ij} - \bar{X}_{.j}) + \varepsilon_{ij} & (7.10)\\
=& \gamma_{00} + \gamma_{01}W_j + \varepsilon_{oj} + (\gamma_{10} + \gamma_{11}W_j + \varepsilon_{1j})(X_{ij} - \bar{X}_{.j}) + \varepsilon_{ij} & (7.11)\\
=& \gamma_{00} + \gamma_{01}W_j + \varepsilon_{oj} + \gamma_{10}(X_{ij} - \bar{X}_{.j}) + \gamma_{11}W_j(X_{ij} - \bar{X}_{.j}) + \varepsilon_{1j}(X_{ij} - \bar{X}_{.j}) + \varepsilon_{ij} \\
& & (7.12)\\
=& \gamma_{00} + \gamma_{01}W_j + \gamma_{10}(X_{ij} - \bar{X}_{.j}) + \gamma_{11}W_j(X_{ij} - \bar{X}_{.j}) + \varepsilon_{oj} + \varepsilon_{1j}(X_{ij} - \bar{X}_{.j}) + \varepsilon_{ij} \\
& & (7.13)
\end{aligned}
$$

The last line just rearranges the random errors to appear at the end of the equation.

Notice that this is no longer a simple linear model: for that to be true the random errors would have to be iid. The random errors have a much more complex structure: $\varepsilon_{oj} + \varepsilon_{1j}(X_{ij} - \bar{X}_{.j}) + \varepsilon_{ij}$ Therefore ordinary least squares will not help us find parameter estimates here.

7.3 The Levels of the Complex Linear Model

The combined model in equation 7.2 is composed of several parts:

The level-1 model:

$$Y_{ij} = \beta_{oj} + \beta_{1j}(X_{ij} - \bar{X}_{.j}) + \varepsilon_{ij} \tag{7.14}$$

The level-2 models:

$$\beta_{oj} = \gamma_{00} + \gamma_{01}W_j + \varepsilon_{oj} \tag{7.15}$$

and

$$\beta_{1j} = \gamma_{10} + \gamma_{11}W_j + \varepsilon_{1j} \tag{7.16}$$

Raudenbush and Bryk (2002) call the β parameters in the Level-1 model the Level-1 coefficients, and γ the Level-2 coefficients.

The above model has a single Level-1 predictor (X_{ij}) and a single Level-2 predictor (W_j). Such a model is called a linear mixed model. Any dataset that has a grouped structure has this structure. In psycholinguistics, within-subject, repeated measures experiments are a good example. Now let's look at how this kind of a model is computed in R. The library we need for linear mixed models was developed by Douglas Bates, and is described in Bates & Sarkar, 2007; see the homepage for lme4:

  http://r-forge.r-project.org/projects/lme4/

For implementation details, see the (currently freely downloadable from the above website) version of Bates' forthcoming book entitled *lme4: Mixed-effects modeling with R*.

```
> lme1.fm <- lmer(MathAch ~ SES + Sector + (1 +
        SES | School), MathScores)
> summary(lme1.fm)

Linear mixed model fit by REML
Formula: MathAch ~ SES + Sector + (1 + SES | School)
   Data: MathScores
   AIC   BIC logLik deviance REMLdev
 46616 46664 -23301    46597    46602
Random effects:
 Groups    Name        Variance Std.Dev. Corr
 School    (Intercept) 3.96385  1.99094
           SES         0.43431  0.65902  0.550
 Residual              36.80088 6.06637
Number of obs: 7185, groups: School, 160

Fixed effects:
            Estimate Std. Error t value
(Intercept)  14.0138     0.2604   53.82
SES           2.3853     0.1179   20.24
```

```
SectorPublic   -2.5409        0.3445     -7.37
```

```
Correlation of Fixed Effects:
            (Intr) SES
SES            0.098
SectorPublc -0.741  0.079
```

The results show that SES and Sector are both significant predictors of math achievement: the higher the SES, the greater the math achievement (specifically, increasing SES by 1 unit increases math achievement by 2.3853 units), and that Public schools have a lower math achievement than Catholic schools (specifically, when the coding for sector is 1, i.e., when the sector is 'Public,' then math achievement is predicted to fall by 2.54 units). Both these results are statistically significant at $\alpha = 0.05$; we know this because the value of the t-scores is greater than 2. The `lmer` function does not deliver p-values; for further details see the note from the author of the `lme4` package, Douglas Bates, at https://stat.ethz.ch/pipermail/r-help/2006-May/094765.html.

We can extract the fixed effect coefficients from the model:

```
> fixef(lme1.fm)
```

```
  (Intercept)             SES SectorPublic
    14.013801        2.385342     -2.540925
```

The estimates for the random effects can also be extracted using the command `ranef(lme1.fm)`.

Let us look at what the variances of the random effects mean:

```
Random effects:
 Groups    Name        Variance Std.Dev. Corr
 School    (Intercept) 3.96385  1.99094
           SES         0.43431  0.65902  0.550
 Residual              36.80088 6.06637
Number of obs: 7185, groups: School, 160
```

1. Var(School)=3.96385 is the variance τ_{00} of the intercept; we saw this earlier for β_{0j}.
2. 0.43431 is the variance of the slopes (τ_{11}).
3. 36.80079 is the variance for the error component $\mathrm{Var}(\varepsilon_{ij})$.
4. 0.550 is the correlation between the slopes and the intercepts, $\mathrm{Cor}(\beta_{0j}, \beta_{1j})$.

We can check whether the correlation between the slopes and intercepts is statistically significant by comparing the current model with a simpler one which does not have a separate slope for each school.

```
> lme0.fm <- lmer(MathAch ~ SES + Sector + (1 |
        School), MathScores)
> summary(lme0.fm)
```

```
Linear mixed model fit by REML
Formula: MathAch ~ SES + Sector + (1 | School)
   Data: MathScores
   AIC   BIC logLik deviance REMLdev
 46621 46656 -23306   46606   46611
Random effects:
 Groups   Name        Variance Std.Dev.
 School   (Intercept)  3.6850  1.9196
 Residual             37.0369  6.0858
Number of obs: 7185, groups: School, 160

Fixed effects:
              Estimate Std. Error t value
(Intercept)    13.8197     0.2527   54.69
SES             2.3748     0.1055   22.51
SectorPublic   -2.1008     0.3411   -6.16

Correlation of Fixed Effects:
            (Intr) SES
SES         -0.066
SectorPublc -0.744  0.091
```

Then, one can compare the two models using analysis of variance:

```
> anova(lme0.fm, lme1.fm)
```

```
Data: MathScores
Models:
lme0.fm: MathAch ~ SES + Sector + (1 | School)
lme1.fm: MathAch ~ SES + Sector + (1 + SES | School)
         Df   AIC   BIC logLik  Chisq Chi Df Pr(>Chisq)
lme0.fm   5 46616 46651 -23303
lme1.fm   7 46611 46660 -23299 9.0387      2    0.01090
```

The more complex model, `lme1.fm`, involves estimating two additional parameters (the variance of the SES slopes by school, and the correlation of the slopes and intercepts) and therefore has two extra degrees of freedom compared to `lme0.fm`. As the ANOVA above shows, the fit of the more complex model is significantly better. This suggests that taking the correlation between the slopes and intercepts into account provides a better model for the data. For much more detail on this and other issues, the reader should consult Baayen, 2008, Baayen, Davidson, & Bates, 2008, Baayen & Milin, 2010, Fox, 2002, Gelman & Hill, 2007, Jaeger, 2008, and Kliegl, Masson, & Richter, 2010. Additional material related to linear mixed models is available from the website of the current book's website.

7.4 Further Reading

If you have read this far, congratulations! This material is not easy, and it will not hurt to work through the book more than once, and to play with the simulations to get a feeling for the ideas we present. This book is intended to prepare you for more advanced textbooks, but of course the really advanced books assume a mathematical background that includes calculus etc.; not everyone will want to go that far. Baayen's is an accessible general textbook involving the use of R for linguistic data analysis, and another excellent text specifically for linear mixed models is Gelman & Hill, 2007. For learning more about R per se, there is a large number of textbooks available, for example the Use R! series published by Springer (e.g., Wickham, 2009 and Sarkar, 2008 for producing high-quality graphics). For a list of other textbooks, see the CRAN website. We particularly recommend Chambers, 2008 for a detailed discussion on programming with R.

Appendix A
Random Variables

This appendix is intended to be a more technical introduction to the concept of random variables. It's optional reading.

Recall the discussion in chapter 3 regarding the raindrops falling on stones. Consider now only four instances of four drops of rain. What is the probability of there being 0 ... 4 Right-stone hits?

Table A.1 The probabilities of 0...4 right-stone hits.

X, the number of R-stone hits	0 1 2 3 4
Probability of R-stone hits	? ? ? ? ?

X above is referred to as RANDOM VARIABLE, and is defined as below:

Definition A.1. A random variable is a real value function defined on a sample space. I.e., $X(e)$ = some real value, e an event in the sample space.

A more general representation of the four-drop scenario is in Table A.2:

Table A.2 A general characterization of the probabilities of $x_1 \ldots x_k$ right-stone hits.

X	x_1 x_2 x_3 ... x_k
Probability	$f(x_1)$ $f(x_2)$ $f(x_3)$ ... $f(x_k)$

The $f(x_1) \ldots f(x_k)$ is the PROBABILITY DISTRIBUTION (contrast this with the FREQUENCY DISTRIBUTION that we've been plotting in past simulations). As an exercise, you may want to think about what the value of the following is: $\sum_{i=1}^{k} f(x_i)$.

S. Vasishth, M. Broe, *The Foundations of Statistics: A Simulation-based Approach*, DOI 10.1007/978-3-642-16313-5,
© Springer-Verlag Berlin Heidelberg 2011

A.1 The Probability Distribution in Statistical Inference

Suppose there are two sentences of English A and B, and we want to know
if participants prefer one or the other in a forced choice test. We start with
the assumption (the null hypothesis) that both are equally preferred. If this
were so, in a random survey of participants, the theoretical probability of one
or the other sentence being preferred is 0.5. So, we can create the *a priori*
probability distribution when, say, 4 participants make a choice (call this one
observation):

Table A.3 The probabilities of A being preferred 0...4 times.

X, the number of A preferences	0	1	2	3	4
Probability	1/16	4/16	6/16	4/16	1/16

Suppose now that we run an experiment using four randomly selected
participants, and we get all participants choosing sentence A. The probability
of this happening is $\frac{1}{16} = 0.0625$. The actual experimental outcome, that all
participants chose A, can either mean that the preference for A is not the
same as the preference for B; or the preferences for A and B are in reality
identical (i.e., the null hypothesis is true), but an unlikely event occurred.

A.2 Expectation

Notice an interesting fact: we can calculate the mean of a bunch of numbers
$x_1, \ldots, x_k$ in two ways. To see this, let's compute the mean of 0,2,2,1,2,3,0,1,2,1.

```
> x <- c(0, 2, 2, 1, 2, 3, 0, 1, 2, 1)
> mean(x)
```

```
[1] 1.4
```

When we use the standard formula for means, we get $\frac{0+2+2+1+2+3+0+1+2+1}{10} =$
1.4 An alternative way is to count the relative frequency of occurrence of each
number, and multiply by that number:

$$0 \times \tfrac{2}{10} + 2 \times \tfrac{4}{10} + 1 \times \tfrac{3}{10} + 3 \times \tfrac{1}{10} = 1.4 = \sum_{i=1}^{k} x_i \times \text{RelativeFrequency}(x_i)$$

That was a computation from a *sample*. Now think of the binomial situation (Heads or Tails, or raindrops falling on stones). Here, the random
variable X can have values 0,1. Suppose we want to know the 'mean' given
the prior probability of a heads $p = 0.5$. Here's how we can compute the
population mean:

$$\mu_X = \sum_{i=1}^{k} (\text{Value} \times \text{Probability}) \qquad (A.1)$$

This population mean is called the EXPECTATION.

Definition A.2. Definition of expectation E(X):

$E(X) = \sum_{i=1}^{k} x_i f(x_i)$

To understand the origin of the term 'expectation,' think of the situation where you were gambling in a casino with a coin, and for each heads you get 50 cents (say, Euro-cents), but you have to pay a playing fee of c Euros for each throw. Then, assuming that the coin is fair, your expected gain is $0 \times 0.5 + 0.5 \times 0.5 = 0.25$ by the above definition. If the casino charges you 1 Euro, the expected gain of the casino in the long run is 50 cents per game.

Note the similarity with the sample mean we just computed using the same formula above, but also note that $\mu = E(X)$ is the population mean here, and is computed from the theoretical probability distribution (which depends on our assumptions about a particular situation like a coin toss), not from any sample.

Consider again the A versus B forced-choice situation.

Table A.4 The probabilities of A being preferred 0...4 times.

X, the number of A preferences	0	1	2	3	4
Probability		1/16	4/16	6/16	4/16 1/16

If we want to know the Expectation of a *function* of a random variable, e.g., $g(X) = (X-2)^2$, there are two ways to do this:

Method 1: Compute distinct values of $(X-2)^2$ and then compute the probabilities of each of these values. This function of the original random variable X is itself a random variable Y now – *but the probabilities associated with Y's values is a function of X's probabilities.* Then apply the definition of Expectation to Y.

- When X=0, $(X-2)^2 = 4$. Probability of X=0: 1/16
- When X=1, $(X-2)^2 = 1$ Probability of X=1: 4/16
- When X=2, $(X-2)^2 = 0$ Probability of X=2: 6/16
- When X=3, $(X-2)^2 = 1$ Probability of X=3: 4/16
- When X=4, $(X-2)^2 = 4$ Probability of X=4: 1/16

Method 2: Compute g(X) and then multiply each with the $f(x_i)$:

Notice that the expectation is the same when computed with the two methods: $\sum = y_i \times f(y_i) = \sum (x_i - 2)^2 f(x_i) = 1$.

The expectation computed by either of the same methods is always going to yield the same result. Reason: In method one we are doing computations

Table A.5 The expectations of a function of a random variable (Method 1).

$Y = (X-2)^2$	0	1	4
$p(Y = y_i)$	3/8	4/8	1/8
$y_i \times p(y_i)$	0	4/8	4/8

Table A.6 The expectations of a function of a random variable (Method 2).

X	0	1	2	3	4
$f(x_i)$	1/16	4/16	6/16	4/16	1/16
$(x_i - 2)^2$	4	1	0	1	4
$(x_i - 2)^2 f(x_i)$	4/16	4/16	0	4/16	4/16

like $g(x) \times (f(x_1) + f(x_3))$ while in method 2 we are doing computations like $g(x) \times f(x_1) + g(x) \times f(x_3)$. These will always yield the same result.

This motivates the way that we define the expectation of a function $g(X)$ of a random variable (so: a function of a function—remember that the random variable is really a function).

Definition A.3. Definition:
$E(g(X)) = \sum g(x_i) f(x_i)$

A.3 Properties of Expectation

Here are some properties of expectations:

(i) $E(a) = a$
(ii) $E(bX) = b \times E(X)$
(iii) $E(X + a) = E(X) + a$
(iv) $E(a + bX) = a + b \times E(X)$
   Proof:

$$E(a + bX) = \sum (a + bx_i) f(x_i) \ldots \text{see above definition of E(g(X))} \quad (A.2)$$
$$= \sum a f(x_i) + \sum bx_i f(x_i) \quad (A.3)$$
$$= a \sum f(x_i) + b \sum x_i f(x_i) \quad (A.4)$$
$$= a \times 1 + bE(X) \ldots \text{because} \sum f(x_i) = 1 \quad (A.5)$$
$$= a + bE(X) \quad (A.6)$$

(v) $E(a + bX + cX^2) = a + b \times E(X) + c \times E(X^2)$
   Proof: exercise.

A.4 Variance

We have worked out so far that $\mu = E(X)$. In order to characterize spread about a mean value, we can use deviations from the mean: $X - \mu$. But this will necessarily give us the answer 0. Suppose $X = x_1, \ldots, x_k$. Then:

$$E(X - \mu) = \sum (x_i - \mu) f(x_i) \tag{A.7}$$
$$= 0 \tag{A.8}$$

So, as before, we square the deviations, and take that as the measure of spread, and call this, as before, Variance:

$$Var(X) = E((X - \mu)^2) \tag{A.9}$$
$$= E(X^2 - 2\mu X + \mu^2) \tag{A.10}$$
$$= E(X^2) - 2\mu E(X) + \mu^2 \ldots \text{from property (v) above} \tag{A.11}$$
$$= E(X^2) - 2\mu^2 + \mu^2 \tag{A.12}$$
$$= E(X^2) - \mu^2 \tag{A.13}$$

And if we scale it down to the dimensions of the mean (as before), we get the standard deviation of the population:

$$sd(X) = \sqrt{Var(X)} = \sqrt{E(X^2) - \mu^2} = \sigma_X \tag{A.14}$$

A.5 Important Properties of Variance

Here are some important properties of variance:

(i) $Var(X+a) = Var(X)$
(ii) $Var(bX) = b^2 Var(X)$
  Proof:

$$Var(bX) = E((bX)^2) - (E(bX))^2 \tag{A.15}$$
$$= E(b^2X^2) - (E(bX))^2 \tag{A.16}$$
$$= b^2E(X^2) - (E(bX))^2 \ldots \text{property (ii) of Expectation} \tag{A.17}$$
$$= b^2E(X^2) - (bE(X))^2 \ldots \text{property (v) of Expectation} \tag{A.18}$$
$$= b^2E(X^2) - b^2E(X)^2 \tag{A.19}$$
$$= b^2(E(X^2) - E(X)^2) \ldots \text{factoring out } b^2 \tag{A.20}$$
$$= b^2(Var(X)) \tag{A.21}$$

A.6 Mean and SD of the Binomial Distribution

At this point we know enough to think about what the mean and standard deviation of the binomial distribution are. Here is the general case of a binomial distribution, Table A.7.

Table A.7 The general case of a binomial distribution.

X	0 1
$f(x_i)$	q p

We can now calculate the variance of X:

$$E(X) = 0 \times q + 1 \times p = p \tag{A.22}$$

$$E(X^2) = 0 \times q + 1 \times p = p \tag{A.23}$$

$$Var(X) = E(X^2) - \mu^2 \tag{A.24}$$
$$= p - p^2 \tag{A.25}$$
$$= p(1 - p) \tag{A.26}$$
$$= pq \tag{A.27}$$

The above is for one observation. For $n > 1$ observations: $X = X_1 + \cdots + X_n$. It follows that (assuming independence of each observation):

$$E(X) = E(X_1 + \cdots + X_n) = E(X_1) + \cdots + E(X_n) = np \tag{A.28}$$

Similarly,

$$Var(X) = E(X_1 + \cdots + X_n) = npq \qquad \text{(A.29)}$$

A.7 Sample versus Population Means and Variances

Assume that the population mean is μ_X and population variance is σ_X. We know that $\bar{X} = \frac{X_1 + \cdots + X_n}{n}$. From the properties of Expectation and Variance discussed above, we can deduce the following two facts:

$$E(\bar{X}) = \frac{1}{n}E(X_1 + \cdots + X_n) \qquad \text{(A.30)}$$

$$= \frac{1}{n}(E(X_1) + \cdots + E(X_n)) \qquad \text{(A.31)}$$

$$= \frac{1}{n}(\mu + \cdots + \mu) \qquad \text{(A.32)}$$

$$= \frac{1}{n}(n \times \mu) \qquad \text{(A.33)}$$

$$= \mu \qquad \text{(A.34)}$$

The above is the sampling distribution of the sampling mean that we discussed at length in chapter 3 (the distribution of the mean values when we repeatedly sample from a population).

This brings us to the relationship between standard deviation and standard error. Consider the variance of $\bar{X}$.

$$Var(\bar{X}) = Var(\frac{1}{n}Var(X_1 + \ldots X_n)) \qquad \text{(A.35)}$$

$$= \frac{1}{n^2}(Var(X_1) + \cdots + Var(X_n)) \qquad \text{(A.36)}$$

$$= \frac{1}{n^2}(\sigma^2 + \cdots + \sigma^2) \qquad \text{(A.37)}$$

$$= \frac{1}{n^2}(n \times \sigma^2) \qquad \text{(A.38)}$$

$$= \frac{\sigma^2}{n} \qquad \text{(A.39)}$$

In other words:

$$\sigma_{\bar{X}} = \sqrt{\frac{\sigma^2}{n}} \qquad \text{(A.40)}$$

This is the standard deviation of the sampling distribution of the sampling means.

A.8 Summing up

Here are some shortcuts we derived in this appendix: To compute mean and
deviation of a sample count X that has binomial distribution $B(n,p)$:

$$\mu_X = n \times p \qquad\qquad\qquad (A.41)$$

$$\sigma_X = \sqrt{n \times p(1-p)} \qquad\qquad (A.42)$$

Suppose we have a population of 1000 students. 600 male, 400 female. We
take one random sample of 40 students. How many females are there?

```
> females <- rbinom(40, 1, 0.4)
> females
 [1] 0 1 1 0 0 1 1 1 1 0 1 1 0 1 0 1 0 1 0 1 0 1 0 1 1 0 0 1
[28] 1 0 0 1 1 0 0 0 0 1 0 1 0

> sum(females)

[1] 21
```

We know that the 95% CI is about $2 \times \sigma_x$. Let's write a function to compute
95 percent CIs for a sample x.

```
> populationsize <- 1000
> samplesize <- 40
> p <- 0.4
> compute95CIpopulation <- function(populationsize,
        samplesize, p) {
        females <- rbinom(samplesize, 1, 0.4)
        samplesumfemales <- sum(females)
        sdsample <- sqrt(samplesize * p * (1 - p))
        sample95CI <- c(samplesumfemales - 2 * sdsample,
            samplesumfemales + 2 * sdsample)
        population95CI <- populationsize/samplesize *
            sample95CI
        print(population95CI)
    }
```

Recall the meaning of the 95% CI: roughly speaking, if we repeatedly take
samples of a given size, 95% of the time the population mean will lie within
it.

Recall that we *know* the population mean here: 40. So, just as an exercise,
sample it 100 times to see what happens. Occasionally (about 5% of the time)
we should get an interval which does not contain the population mean.

```
> compute95CIpopulation(1000, 40, 0.4)

[1] 370.0807 679.9193
```

Problems

A.1. Prove the following statement:
$$E(a + bX + cX^2) = a + b \times E(X) + c \times E(X^2)$$

Appendix B
Basic R Commands and Data Structures

This concise introduction to R commands and data structures illustrates the basic level of knowledge required to understand the code used in the book.

We can multiply (divide, etc.) numbers by numbers, which gives a new number:

```
> 10 * 3
```

```
[1] 30
```

We can *assign* numbers to *variables*, and access their values by invoking the variable:

```
> n <- 10
```

```
[1] 10
```

We can create *vectors*, which are lists of numbers:

```
> v <- c(1, 15, 4, 27, 5)
```

```
[1]  1 15  4 27  5
```

There are shortcuts for creating vectors with simple patterns:

```
> v <- 1:5
```

```
[1] 1 2 3 4 5
```

We can multiply (divide, etc.) vectors by numbers, which gives a new vector:

```
> w <- v * 3
```

```
[1]  3  6  9 12 15
```

Some *functions* work element-by-element as above, like the function abs():

```
> abs(-2:2)
```

```
[1] 2 1 0 1 2
```

We can get help for any function by preceding it by ?:

```
?abs
```

Some functions take all the inputs and return a single number, like the function sum():

```
> sum(-2:2)
```

```
[1] 0
```

The individual numbers in a vector can be accessed using their *positional index*:

```
> w[3]
```

```
[1] 9
```

We can access each number in a vector in turn using a for-loop and a variable index:

```
> for (i in 1:5) {
      print(w[i])
  }
```

```
[1] 3
[1] 6
[1] 9
[1] 12
[1] 15
```

We can put the same number into the cells of a vector, using rep() (repeat):

```
> v <- rep(3, 5)
```

```
[1] 3 3 3 3 3
```

We can put numbers into a vector using a positional index, as long as we initialize the vector first; usually this initialization is done using empty slots filled with NA's:

```
> v <- rep(NA, 5)
> for (i in 1:5) {
      v[i] <- i * 3
  }
```

We can do various things to a number before we finally put it in a vector:

```
> v <- rep(NA, 5)
> print(v)
```

```
[1] NA NA NA NA NA
> for (i in 1:5) {
      square <- i^2
      v[i] <- square - 1
  }
> print(v)

[1]  0  3  8 15 24
```

A *matrix* is a two dimensional table of numbers, with (vertical) columns, and (horizontal) rows:

```
> m <- matrix(1:6, ncol = 2, nrow = 3)

     [,1] [,2]
[1,]   1    4
[2,]   2    5
[3,]   3    6
```

We can access the individual numbers in a matrix using a two dimensional index:

```
> m[3, 2]

[1] 6
```

An entire column (or row) can be accessed:

```
> m[, 2]

[1] 4 5 6

> m[2, ]

[1] 2 5
```

These columns (rows) behave just like vectors:

```
> m[, 2] * 3

[1] 12 15 18
```

Entire matrices can be manipulated like vectors:

```
> m * 3

     [,1] [,2]
[1,]   3   12
[2,]   6   15
[3,]   9   18

> sum(m)
```

`[1] 21`

We can access each value in a matrix in turn using nested `for`-loops: `for` column 1, `for` each row...; `for` column 2, `for` each row...; etc.

```
> for (j in 1:2) {
      for (i in 1:3) {
          print(m[i, j])
      }
   }
```

```
[1] 1
[1] 2
[1] 3
[1] 4
[1] 5
[1] 6
```

A `data.frame` is a matrix whose columns are labeled, analogous to a spreadsheet:

```
> d <- data.frame(smaller = 1:3, larger = 4:6)
```

```
  smaller larger
1       1      4
2       2      5
3       3      6
```

We can access each column in a data.frame using either a positional index, or (more commonly) its label:

```
> d[[1]]
```

`[1] 1 2 3`

```
> d$smaller
```

`[1] 1 2 3`

References

Baayen, R. (2008). *Analyzing Linguistic Data. A Practical Introduction to Statistics Using R.* Cambridge University Press.

Baayen, R., Davidson, D., & Bates, D. (2008). Mixed-effects modeling with crossed random effects for subjects and items. *Journal of Memory and Language, 59*(4), 390–412.

Baayen, R., & Milin, P. (2010). *Analyzing Reaction Times.* (To appear in International Journal of Psychological Research)

Bates, D., & Sarkar, D. (2007). *lme4: Linear mixed-effects models using s4 classes.* (R package version 0.9975-11)

Chambers, J. (2008). *Software for data analysis: Programming with R.* Springer Verlag.

Cohen, J. (1988). *Statistical power analysis for the behavioral sciences* (2 ed.). Hillsdale, NJ: Lawrence Erlbaum.

Fox, J. (2002). *An R and S-Plus companion to applied regression.* Sage Publications, Inc.

Gelman, A., & Hill, J. (2007). *Data analysis using regression and multilevel/hierarchical models.* Cambridge, UK: Cambridge University Press.

Hoenig, J. M., & Heisey, D. M. (2001). The abuse of power: The pervasive fallacy of power calculations for data analysis. *The American Statistician, 55*(1), 19-24.

Jaeger, T. (2008). Categorical data analysis: Away from ANOVAs (transformation or not) and towards logit mixed models. *Journal of Memory and Language.*

Kliegl, R., Masson, M., & Richter, E. (2010). A linear mixed model analysis of masked repetition priming. *Visual Cognition, 18*(5), 655–681.

Leisch, F. (2002). Sweave: Dynamic generation of statistical reports using literate data analysis. In W. Härdle & B. Rönz (Eds.), *Compstat 2002 — proceedings in computational statistics* (pp. 575–580). Physica Verlag, Heidelberg. (ISBN 3-7908-1517-9)

Maxwell, S. E., & Delaney, H. D. (2000). *Designing experiments and analyzing data.* Mahwah, New Jersey: Lawrence Erlbaum Associates.

Oakes, M. (1987). *Statistical inference: A commentary for the Social and Behavioral Sciences*. NY: John Wiley and Sons.

Polya, G. (1954). *Mathematics and Plausible Reasoning, vol. 2*. Princeton, NJ: Princeton University.

Raudenbush, S. W., & Bryk, A. S. (2002). *Hierarchical linear models: Applications and data analysis methods* (Second ed.). Sage.

Rice, J. (1995). *Mathematical statistics and data analysis*. Duxbury press Belmont, CA.

Rietveld, T., & van Hout, R. (2005). *Statistics and language research: Analysis of variance*. Berlin: Mouton de Gruyter.

Rosen, K. H. (2006). *Discrete mathematics and its applications* (Sixth Edition ed.). New York: Mc-Graw Hill, Inc.

Sarkar, D. (2008). *Lattice: Multivariate data visualization with R*. Springer Verlag.

Schuirmann, D. (1987). A comparison of the two one-sided tests procedure and the power approach for assessing the equivalence of average bioavailability. *Journal of Pharmacokinetics and Pharmacodynamics, 15*(6), 657–680.

Stegner, B. L., Bostrom, A. G., & Greenfield, T. K. (1996). Equivalence testing for use in psychosocial and services research: An introduction with examples. *Evaluation and Program Planning, 19*(3), 193-198.

Student. (1908). The probable error of a mean. *Biometrika, 6*(1), 1–25.

Walker, H. (1940). Degrees of freedom. *Journal of Educational Psychology, 31*(4), 253–269.

Wickham, H. (2009). *ggplot2: Elegant graphics for data analysis*. Springer Verlag.

Index